Tobias Haller, Claudia Zingerli (eds.)
Towards Shared Research

Culture and Social Practice

Tobias Haller (Prof. Dr.), born 1965, is a professor at the Institute of Social Anthropology at the University of Bern, Switzerland, with a focus on economic, ecological, and political anthropology.
Claudia Zingerli (PhD), born 1973, is a scientific collaborator at the Swiss National Science Foundation and freelancer in diverse activities dealing with knowledge brokerage and the co-creation of knowledge in science-policy interfaces.

Tobias Haller, Claudia Zingerli (eds.)
Towards Shared Research
Participatory and Integrative Approaches
in Researching African Environments

[transcript]

This book was possible because of the perseverance of the authors and the financial support from the Swiss Academy of Humanities and Social Sciences, the Swiss Academy of Sciences, the Swiss Academic Society for Environmental Research and Ecology (saguf) and the Swiss Society for African Studies (SSAS).

Bibliographic information published by the Deutsche Nationalbibliothek
The Deutsche Nationalbibliothek lists this publication in the Deutsche Nationalbibliografie; detailed bibliographic data are available in the Internet at http://dnb.d-nb.de

This work is licensed under the Creative Commons Attribution 4.0 (BY) license, which means that the text may be be remixed, transformed and built upon and be copied and redistributed in any medium or format even commercially, provided credit is given to the author. For details go to http://creativecommons.org/licenses/by/4.0/
Creative Commons license terms for re-use do not apply to any content (such as graphs, figures, photos, excerpts, etc.) not original to the Open Access publication and further permission may be required from the rights holder. The obligation to research and clear permission lies solely with the party re-using the material.

First published in 2020 by transcript Verlag, Bielefeld
© Tobias Haller, Claudia Zingerli (eds.)

Cover layout: Maria Arndt, Bielefeld
Cover illustration: Women engaging in shared research regarding the development and testing of cook stoves at the Great African Cook-Off in Malawi, one of the examples in the book on how participatory research is carried out successfully (see paper by Jewitt et al., © picture: Charlotte Ray and Maria Beard).
Printed by Majuskel Medienproduktion GmbH, Wetzlar
Print-ISBN 978-3-8376-5150-8
PDF-ISBN 978-3-8394-5150-2
https://doi.org/10.14361/9783839451502

Contents

Foreword 9

Towards collaborative and integrative research in African environments
An introduction
Tobias Haller and Claudia Zingerli 11
1.1. African environments in focus 11
1.2. Spiralling (mis)interpretations 12
1.3. Fragmented knowledge 14
1.4. Longitudinal knowledge guidance for researching African environments today .. 16
1.5. Towards shared research 18
1.6. Overview of contributions 19
1.7. References 21

Soil classifications
Between material facts and socio-ecological narratives
Brice Prudat, Lena Bloemertz, Olivier Graefe, Nikolaus Kuhn 25
2.1. Introduction 25
    2.1.1. Ohangwena region and villages 26
    2.1.2. Collecting local soil knowledge 27
    2.1.3. Scientific soil description 28
    2.1.4. The Oshikwanyama soil units 29
    2.1.5. Local soil types compared to international classifications 30
    2.1.6. Advantages of combining local and scientific knowledges 32
2.2. Issues regarding the participatory approach in natural sciences 33
    2.2.1. Translations of the concept of "soil" 33
    2.2.2. Intergrades 34

  2.2.3. Local experts 35
  2.2.4. Accuracy of descriptions... 35
 2.3. Participatory research in natural sciences: reflections and challenges 39
  2.3.1. Expectations and managing data 39
  2.3.2. Dealing with complexity 40
 2.4. Conclusion and perspectives... 41
 2.5. References 43

Action research and reverse thinking for anti-desertification methods
Applying local revegetation techniques based on the ecological knowledge of local farmers in the Sahel of West Africa
Shuichi Oyama 47
3.1. Introduction...47
3.2. Desertification in the Sahel region... 48
3.3. Approach and research area 50
3.4. Agriculture in long-term dry season and short rainy season 52
 3.4.1. Temperature, rainfall and wind 52
 3.4.2. Agriculture 54
 3.4.3. Soil properties and land degradation 55
3.5. Local countermeasures against land degradation 59
 3.5.1. "Waste is manure for our farmland".. 61
 3.5.2. First trial of urban waste-induced land restoration 63
 3.5.3. Emerging pastureland 66
3.6. Eight effects of urban waste use for land restoration................................... 68
 3.6.1. Safety issues with urban waste 72
 3.6.2. Collecting waste from the city administration to resolve the financial deficit problem 73
 3.6.3. Inviting livestock into the fenced pastureland 77
3.7. Conflict prevention and livestock-induced land restoration 79
3.8. Conclusion: urban waste, new institution and combating desertification 83
3.9. References 86

Energy and the environment in Sub-Saharan Africa
Household perceptions of improved cookstoves
Sarah Jewitt, Peter Atagher, Mike Clifford, Charlotte Ray and Temilade Sesan 91
4.1 Introduction... 91
 4.1.1. The evolution of improved cookstove initiatives................................. 91

    4.1.2. Recent initiatives promoting clean fuels and cookstoves 92
    4.1.3. Neglect of end-user preferences 94
    4.1.4. Limitations of fuel and ICS monitoring 95
    4.1.5. Research problem and contribution 96
4.2. Methodological approaches...96
    4.2.1. Bake/cook-off events 98
    4.2.2. Field-based research in Benue State 99
    4.2.3. Field-based methodologies 101
4.3. End-user priorities for cooking systems: results from the bake/cook-off events 103
4.4. Community-level perspectives on cooking systems and fuel choices in Benue . 108
    4.4.1. Class and gender as influences on ICS and fuel use 108
    4.4.2. Access to firewood 109
    4.4.3. Smoke-related concerns versus household budget constraints 110
    4.4.4. Socio-cultural factors influencing stove and fuel stacking 111
    4.4.5. User preferences for rapid cooking 114
    4.4.6. Seasonal shifts in stove and fuel use.. 114
4.5. Incorporating end-user preferences into stove interventions and SDG7 monitoring frameworks.. 115
4.6. References 119

Fishing for food and food for fish
Negotiating long-term, sustainable food and water resources in a transdisciplinary research project in Burkina Faso
Gabriele Slezak, Jan Sendzimir, Raymond Ouedraogo, Paul Meulenbroek, Moumini Savadogo, Colette Kabore, Adama Oueda, Patrice Toe, Henri Zerbo and Andreas Melcher 125
5.1 Research context 125
    5.1.1. The establishment of a transdisciplinary research project 126
    5.1.2. Integrating practices of participatory research 127
    5.1.3. Project results 130
    5.1.4 Issues with the participatory approach 132
    5.1.5. Fieldwork - practice and training 133
    5.1.6. Involvement of policy makers - key questions of management............. 139
    5.1.7. Synthesis of research results... 141
    5.1.8. SUSFISH's participatory approach: lessons learned and problems 147
5.2. Key moments of participatory research... 150
    5.2.1 Scenario development workshops - key to understanding 150

5.2.2. The debate is open: translational practices to negotiate meaning	153
5.2.3. The debate on gender	156
5.3. Conclusion and main learnings	159
5.4. References	162

Conclusion
Explorations and lessons for shared research
Claudia Zingerli and Tobias Haller 167

6.1.	Explorations	167
6.2.	Learning as a multidimensional and multilevel process	169
6.3.	Dimensions of participatory research	172
6.4.	Role of language and translation in interdisciplinary and intercultural research settings	173
6.5.	Turning points in collaborative research processes	175
6.6.	Towards shared research	176
6.7.	References	178

Authors 181

Foreword

In October 2015, a diverse group of scholars of different disciplinary (humanities, social and natural sciences) and geographical origins and at different academic career stages met for a conference on "Participatory and Integrative Approaches in Researching African Environments: Opportunities, Challenges, Actualities in Natural and Social Sciences". All of them came with an open multidisciplinary perspective and portrayed in fascinatingly transparent ways their approaches and searches for a more inclusive and better understanding of knowledge about African environments. Some presenters focused on their methodological attempts to structure problem framings and knowledge production processes in transdisciplinary ways, including not only scientists but also stakeholders from local contexts of environmental resource use, administration and policy. Some gave insights into critical reflections on global policy frameworks, categorizations and interpretations and their (mis)fits with local realities, highlighting the diversity of knowledge and techniques for dealing with environmental and social change.

Four points emerged from the conference regarding the process of what we called shared research.

- Unintended and unanticipated chances and challenges emerge in collaborative (participatory and integrative) research on African environments. Being open to and respectful of local level actors enhances learning and joint contributions to facilitate a common understanding of key aspects in researching African environments.
- Reflecting critically on research processes and multiple expectations by formulating (in writing or speech) deeper insights and discomfort affects the dimensions of learning. A change of perspective on the research process can shift paradigms and epistemological traditions.
- Doing participatory and integrative research on African environments means being explicit about research ethics and the role of those who get

in and out of a local context. For more inclusive learning, mutual respect and trust are key.
- Transdisciplinary research needs careful planning of participation and an openness towards emerging participation while the research progresses.

All in all, it was rich learning experience with conference participants who dared to talk about their struggles with language and interpretation and science as contribution, which is often not possible in disciplinary conferences. Participants facilitated a better understanding of complex socio-political and environmental systems. The conference participants enjoyed the contributions and discussions to the extent that the idea of further elaborating on some of the topics emerged. In the form of an edited volume, which provides ample room for each contribution to go into detail, the authors present their contributions in explicit and nuanced ways.

Finally, the volume is put together as a collection of four articles dealing with similar challenges, opportunities and actualities. The authors were given a lot of room for exploration during the writing process, from which four heterogeneous papers emerged. In the introductory and concluding articles, the editors highlight and provide a discursive framework in which the articles can be positioned. They discuss key issues for an audience interested in reflections and twists in relation to participatory and integrative research in intercultural, interdisciplinary and transdisciplinary interfaces in selected African environmental contexts.

The editors are very grateful to the Swiss Academy of Humanities and Social Sciences for financial support for both the conference and the publication of this book, particularly the Open Access publication grant. We also highly acknowledge the financial contributions of the Swiss Academy of Sciences, the Swiss Academic Society for Environmental Research and Ecology (SAGUF), and the Swiss Society for African Studies (SSAS). We are indebted to Samuel Weissman, who supported the process in critical stages.

Bern, January 2020

Tobias Haller and Claudia Zingerli

1. Towards collaborative and integrative research in African environments
An introduction

Tobias Haller and Claudia Zingerli

1.1. African environments in focus

The legacy of the colonial gaze at African environments has been an issue in critical studies in the humanities and some natural sciences for more than 30 years now. Fuelled by the emerging political ecology approach (see Robbins 2004), research and publications have radically challenged the way the postcolonial world views African environments. One of this view's cornerstones was the award-winning publication *Misreading the African landscape* (Fairhead and Leach 1996), in which it became evident that forest patches in the Guinean Savannah were not pristine remains of a large forest cover destroyed by local people. On the contrary, the authors showed that forest patches were planted and thus created by local people for multiple purposes. This new paradigm that so-called natural environments are not purely natural but cultural landscaped ecosystems has been an important narrative in ecological anthropology, with its roots in the 1980s and 1990s (see Roy Ellen's 1982 milestone book on subsistence production or Netting's *Smallholders, householders* in 1993). However, this paradigm shift remained bound in the sub-discipline and did not extend into other disciplines or even interdisciplinary arenas.

Fairhead and Leach's 1996 book was vital for a paradigmatic and discursive shift. It marked a wider recognition in the interdisciplinary and transdisciplinary worlds, because it combined sound social anthropological qualitative research with historical archive research, and more quantitative data stemming from air photography, including digital satellite images from geographic and remote sensing. This mixed methods approach established the basis for data, contributing to a wider recognition of the paradigm shift and challenging previously held views - views that were based on negative labelling of people in African environments as trapped in a tragedy of environmental degradation; views that were very deeply rooted in the colonial discourse regarding the need to protect forests and wildlife from damage caused by the overuse of natural resources by local people (see Haller and Galvin 2011, Neumann 1998); views that called for the preservation of "pure" nature, which had to be carried out by outside "civilized" actors and was part of colonial legitimacy, which is carried on today among many conservationist views (see Brockington et al. 2008, Galvin and Haller 2008).

For looking at African environments, their reading through scientific, societal and political lenses, and interpretations about their emergence and use, the merit of the book lies not only in its mixed methodology; it lies in the integrative turn in social anthropological research which created room for the way research was conducted on the ground. Fairhead and Leach (1996) were very open to different local views and conceptualizations of the so-called environment in their research and to local explanations of why the forests looked like they did. From that emerged a more participatory research agenda, being open to the way local people viewed the environment and developed cultivation strategies for these forests. Without such an approach, these new scientifically ground-breaking insights could not have been discovered. The authors learned about how people perceived themselves symbolically and made analogies to different animals such as termites in the environment, while at the same time indicating techniques and reasons for forests ranging from shade to economic, religious and political reasons (such as defence). These views were embedded in spiritual worldviews, which placed local people in a world interacting with spirits and ancestors, and so being aware from their own animistic religious perspective (the emic view) that they are not alone in this cultural landscape ecosystem but that there is a need to interact and communicate ritually with this spiritual world (see Haller 2019a, 2019b).

1.2. Spiralling (mis)interpretations

The colonial and post-colonial conservation gaze on the environment and on people living in that cultural landscape ecosystem was not simply misread but was coupled with an incomplete and biased analysis, which, however, became politically important. The idea that drought in the area was caused by local clear-cutting of an anticipated full pristine forest covering the whole zone led to repressive colonial and post-colonial policies and the labelling of people as forest or savannah people. It is part of what James Scott (1998) called "seeing like a state", by which he meant numbering, standardizing and labelling the environment and its people for policy action and control. This process created, and still creates, rules which are not only not adapted to local contexts, but that also led to political subordination and, as a consequence, to more degradation, creating what is called a "positive feedback loop" in system theory. The reinforcing of wrongly labelled pristine nature in peril led to repressive policies which, as a reaction from local stakeholders, leads to destructive counter-reactions as property rights are taken out of the hands of local people. When local people lose their common property rights to land and land-related resources as a form of colonial and post-colonial commons grabbing, they lose their sense of ownership and belonging. The reaction is like the one described by Hardin (1968), yet not as a tragedy of the commons but as a tragedy of the grabbed commons, leading to state control and, as states are not efficient at controlling, to the tragedy of open access. Cases in Africa such as in Guinea, Tanzania, Cameroon, Zimbabwe and Zambia show that local people rather destroy forests and wildlife under these conditions of grabbing before others take the resources, and without wildlife present, for example, conservationists will leave the land to the people as there is no longer any reason for protection (see Murphree 2001, Haller 2010, 2016).

Such counter-reactions unfortunately can then again be taken by conservationists as proof that Africans only see charcoal and game meat for the pot behind forests and wildlife, reinforcing a process of fortress or top-down so-called participatory approaches to conservation (see Galvin and Haller 2008, Haller et al. 2016). Thus, the African environment and the people living in and from it deserve adequate research and analysis that is translated into adequate policies. This is then a political process to be addressed. We need not be naïve and think that the paradigm shift will change all, as this view produced gains and obligations since colonial times which could not and still cannot today be challenged quickly. We would have to focus on authoritative power and considerable financial implications, and internationalized and interdependent spheres of environmental policy and economics.

1.3. Fragmented knowledge

Regarding the paradigm shift, it needs to be highlighted that the main problem with the ideology of pure nature is the separation of nature and culture (see Descola 2013). This separation has been a main feature of Cartesian thinking and logical reasoning since the time of enlightenment. We do not argue here that insights in that historic period are necessarily wrong, but that they led to the hegemonic view that modernity produces scientifically objective knowledge, while all older and other forms of knowledge and views are seen to be backward and tied to a dark age. Previous or other societies are labelled "traditional", while other ways of looking at what we call the environment are overlooked and their heritage remains largely unaddressed, without history and knowledge.

This view also begs a historical and environmental reality check. Wherever one studies the state of a common pool of resources - irrespective of how bad the figures and numbers might be - one must acknowledge that the biggest loss of landscapes and biodiversity occurred *after* and not before the age of enlightenment and the ages of colonial expansion. This basically means that the biodiversity which is lost now was there before colonialization with the presence of local peoples - and the work of Fairhead and Leach (1996) and others suggests that this biodiversity existed because of local people's management of resources.

Around the globe, cultural landscape ecosystems have been developed over centuries based on the views, regulations (institutions) and uses (practices) of local peoples. The term "institution" refers to formal and informal dos and don'ts, or to the rules of the game (see North 1990), as well as property rules, regulations of use, norms and practice-induced values (see Ensminger 1992, Haller 2010). These institutions are also full of condensed knowledge and serve as an orientation for collective action to maintain cultural landscapes based on the wisdom that interactions with the environment (the cultural landscape ecosystem context) need coordinated action with other humans and other groups of humans (the political environment) as well as with the world of spirits and souls of the material and immaterial and ancestors (the spiritual environment) (Haller 2007, 2010). Not taking this knowledge into account during colonial and post-colonial times, however, did not only reduce sociocultural aspects; it also had a negative impact on the environment on several levels. It reduced knowledge on how to deal with the created landscapes and how to maintain and manage them (see Bornemann et al. 2017). A reduced knowledge base results in degradation of these cultural (and natural) landscapes.

Therefore, there is a claim to recognize knowledge among local groups. Alexander von Humboldt (1852, Lubrich 2009) was not so much the one discovering and "inventing nature" in the Americas - as Wulf (2016) makes us believe - but the one discovering complex interrelationships between different factors such as climate, topography, vegetation and soils, as well as animals and humans transforming these environments. On his journeys through Latin America, Humboldt realized that landscapes had been inhabited and that colonialization left massive tracts in the landscapes and altered them. He was therefore interested in the views of local people in the same way as he was interested in the data gathered by his instruments and the information he received from other scholars. He realized before the debate on the nature-culture divide in humanities and natural sciences that local indigenous groups did not make this division, a fact highlighted later on by scholars such as Tim Ingold (2000) and explicitly by Philippe Descola (2013). Their emphasis, especially by Descola, on animistic and totemistic worldviews, characteristic of local peoples' environmental worldview, was to show the notion local groups had of not being alone but rather embedded in a much larger environment. The result of that relationship is an emic view of mutual interaction between the material and the spiritual.

On the other hand, the Cartesian separation on which natural science is based created a new construction of nature as not influenced by humans. Humboldt was exceptional as he was not guided by this divide in his scientific curiosity but thought in terms of interrelations and was interested in different views. On the one hand, he focused on natural science methodology and gathered quantitative data, but on the other hand, he was interested in the views and knowledge other peoples had on how the environment functioned. In addition, he paid attention to their practices and rationales by gathering information on their views on processes in the environment, which he considered to be information as important as the natural science data. What is unclear with Humboldt is the question of whether he considered local people as creating cultural landscape ecosystems. However, on his journeys throughout the Americas he recognized the infrastructure created by the Incas as equal to the infrastructure created by the Roman Empire, and he explained environmental changes such as low water levels of lakes as stemming from the colonial practice of plantation economy, and thus the altering of forest cover as a major cause. Similarly, he described the use of guano (a natural fertilizer stemming from bird dung on islands) as being sustainably managed by the Incas, which means he knew the impact local peoples had on a resource that can be overused if not managed carefully (Humboldt 1853). Therefore, we argue that he was also attentive to the historical processes of altering of the landscape by humans, both indigenous and immigrants through the colonial process, and he criticized the latter for their unsustainable resource management.

1.4. Longitudinal knowledge guidance for researching African environments today

By allowing plural views and putting them into historical contexts, Humboldt was the first scholar to use what we call a participatory research orientation. This orientation is the basis of Fikret Berkes' (1999) work, who gave sense to Humboldt's thinking without actually referring to Humboldt. Berkes tried to show the differences and similarities between the scientific and local indigenous peoples' thinking, and reached the following conclusions by studying the knowledge systems of First Nation peoples in the US and in Canada. The basic difference between the natural scientific and local indigenous knowledge is that the latter is not expert knowledge: it is not developed by someone distant, but is locally embedded, and it is related to practice and transmitted through generations. It thus has what natural science knowledge often lacks: a longitudinal basis of accumulated knowledge that can also be very adaptive through time. The basic similarities between natural science ecosystem knowledge and indigenous ecological knowledge can be seen in the way both knowledge systems try to reduce complexity and are interested in experimenting. By this, Berkes - like other scholars, such as Paul Richards (1985) - means being engaged in processes of trial and error in the field, for example in agriculture, by trying out different crop varieties, irrigation patterns, etc. Both knowledge systems deal with practical notions, but the indigenous system focuses centrally on the issue of risk management to secure livelihoods, as some authors have shown in relation to hunter-gatherer societies (Sahlins 1972, Winterhalder and Smith, 1985, and many others), pastoral societies (Homewood 2008, McCabe 1990), agriculturalists (Ellen 1982, Netting 1993, Richards 1985), and many mixed forms (hunting and horticulture, agro-fishery-pastoralism, fishing-agriculturalists), as well as resource-based occupational groups being interrelated (arrangements between farmers, fisheries and nomadic pastoralists) (for a summary, see Haller 2007, see also Haller 2010, 2013, 2016, Haller et al. 2013).

There is a lot of accumulated practical knowledge handed down over generations and adapted to local changing conditions in these societies, which basically work on the principle of longitudinal knowledge of dynamics and expected variability in the cultural landscape ecosystems. The basic common strategy is not to drive for the best but for maximization of the minimal needed yields, catches, game hunted etc. Such a mini-max strategy (see Haller 2007) can only work by diversifying uses and accumulating generational knowledge on environmental dynamics in cultural landscape ecosystems. Complementarily, local peoples have developed worldviews which help to explain uncertainty and often provide ritual practices to bring (from an emic perspective) an unbalanced human-spiritual world relationship back into balance. These elements - knowledge, resource management and practices on the one side, as well as worldviews - are interconnected on the level of what Berkes (1999) calls social institutions.

Where this works, the results of adapting to changes such as economic and climatic changes produce better solutions on several levels. Firstly, local actors have a more longitudinal knowledge of their environments, and interacting on a shared level with scientists can only be beneficial for both sides if science shows more openness. Secondly, adapted rules based on older solutions and knowledge reduce transaction costs in a tremendous way, as feeling a sense of ownership of the knowledge production and crafting processes creates among local people the feeling that the new institutions are theirs and derive from their knowledge. This embodiment of a sense of ownership of the institution-building process, which has been labelled "constitutionality" (as a counter-position to Foucault's "governmentality" and Arun Agrawal's "environmentality") as a collective conscious way of creating institutions (see Haller et al. 2016, 2018) shows very positive effects in a new more sustainable way for resource management. As new literature on Africa shows (Chabwela and Haller 2010, Faye et al. 2018, Haller et al. 2013, Haller and Merten 2008, 2018), supported for other places in the world (see the 2018 special issue of *Human Ecology*, volume 46, issue 1), a process of shared research is needed to develop locally rooted and sustainable institutions.

Therefore, shared research is a central step to such institution crafting, and a process where we still need to learn more. In the history of anthropology, we find such approaches in what was called action anthropology (since 1950, but discontinued; see the work of Sol Tax in Foley 1999) and applied anthropology (from 1980 onwards; see Bennett 1996, Rylko-Bauer et al. 2006). Researching African environmental contexts with explicitly inclusive lenses regarding co-researching and co-learning can fuel this learning process.

1.5. Towards shared research

When organizing the conference "Towards Shared Research: Participatory and Integrative Approaches in Researching African Environments: Opportunities, Challenges, Actualities in Natural and Social Sciences" in 2015, we hoped to profit from rare cases of in-depth insights into the research process and the diversity of knowledge and perspectives about African environments. The conference participants openly debated and challenged their shared research experiences, wherein lies a learning potential with paradigmatic dimensions. All the papers included in this book stem from a deep reflection on and a curiosity in local contexts, but they also show the challenges and turning points in interdisciplinary research and local people-researcher interactions. Researchers in this edited volume have different social and natural sciences and engineering backgrounds, in which an interest in the practice of the "other" and challenges were central to reaching more collaboration and mutual learning. The learning in researching African environments happens on at least two levels: 1) among disciplines of different science domains; and 2) between researchers and the researched. Co-experimenting at these two levels is something which we see in all the papers. It allows reflection on political-historical and power-specific contexts and enables a better understanding of other positions and views (see Zingerli 2010). That again facilitates the notion of co-learning as a basis for shared research. Coming back to the contributions by Fairhead and Leach (1996), shared research might continue to trigger paradigmatic shifts, indicating that numerous variables and different views are of value and shape the way we perceive what is important about the environment and development in African environments today.

For our interpretation of shared research based on the contributions united in this book, we used a framework according to which the presentations can be positioned. It includes the following four elements:

- Learning as multidimensional and multilevel processes in extended time and scale.

- Dimensions of participatory research.
- Role of language and translation in interdisciplinary and intercultural research settings.
- Turning points in collaborative research processes.

In the last chapter, we address each of the elements as a stimulus for further thought and exploration towards collaborative and integrative research in African environments.

1.6. Overview of contributions

The following four original contributions are grouped under two headings: 1) Contextualizing soil fertility; and 2) Negotiating knowledge and technological inventions in intercultural settings.

The two articles by Prudat et al. ("Soil classifications: Between material facts and socio-ecological narratives") and Oyama ("Action research and reverse thinking for anti-desertification methods") both take a focus on soil, soil fertility and soil management in arid sub-Saharan African environments. Prudat et al.'s geographical context is north-central Namibia, while Oyama's study sites are in south-west Niger.

Prudat et al. set out in 2014 to compare local knowledge on soils with two international soil classification systems. They had designed their study from a natural science perspective, focusing on describing soils in a scientific way. The long fieldwork enabled them to delve deep into the complexities of local knowledge about soils and to reflect thoroughly on what it means to give justice to the diversity of local perspectives and to make use of complementary knowledge of soil and soil management. The paper shows what a shared research offers, acknowledging soil as part of a human-made landscape and soil knowledge as part of a socially constructed knowledge. On the other hand, the emic, i.e. local way of understanding soil characteristics has limits that an objectivized soil classification can counterbalance in soil management decision-making. Prudat et al. offer an honest reflection on environmental scientists that enhance their natural sciences' "socialization" through participatory research methods and observation among the Oshikwanyama in north-central Namibia.

Local knowledge of soil and soil management is also a focus for Oyama, who draws on a series of research stays in south-western Niger spanning more than 15 years of participatory research. Oyama's contribution evolves in the context of combating desertification, but it draws attention to wicked problems that manifest in local settings, including the pressure for farmland that reduces pastureland, with ever more livestock and lingering social conflicts. His contribution to enhancing soil fertility and land management can be read as a quest to mitigate a situation of environmental and socio-political stress. Because the author demonstrates high engagement in actual fieldwork and participatory observation, he creates social relations to experiment with and scientifically measures local techniques for improving soil conditions. What he refers to as "reverse thinking" is to apply and test a locally emerging soil management technique for its potential to create plots for enhanced soil fertility and more productive biomass production for livestock herding. What is counter-intuitive from an environmental point of view is the application of solid waste from the city to abandoned and degraded soils. Oyama portrays positive effects on both soil fertility and land management between Hausa farmers and Fulbe herders. Despite large scale programmes to combat desertification in the region, local communities organize themselves for higher productivity. Oyama's action research approach is an interesting and puzzling contribution to dealing with a wicked problem.

The two contributions by Jewitt et al. ("Energy and the environment in sub-Saharan Africa: Household perceptions of improved cookstoves") and Slezak et al. ("Fishing for food and food for fish: Negotiating long term, sustainable food and water resources in a transdisciplinary research project in Burkina Faso") emerge from European-African research collaborations in which the negotiation of knowledge and technological inventions in intercultural settings plays a big role. The paper by Jewitt et al. presents an excellent opportunity to discuss the development of technical innovations based on a health-environment-technical approach. The article explores the evolution of improved cookstove initiatives and looks at initiatives promoting clean fuels and cookstoves. Its analysis evolves against the background of specifically designed events called "bake/cook-offs". The international group of researchers organized cooking events in three largely different settings in order to collect end-users' views. The first bake-off took place in England, with the participation of immigrants and refugees; later, the bake/cook-offs were organized in Malawi and Zambia as well as in Benue state, Nigeria, the latest accompanied by fieldwork. This participatory approach to experimenting and testing makes contributions beyond the health-environment-technology approach. It shows the potential of end-users' views and cultural considerations in processes aimed at introducing new or alternative technologies. It is helpful in understanding the level of adoption of a technology and the power of co-creation of knowledge.

Co-creation of knowledge is a key focus of Slezak et al.'s contribution. It provides a thick account of shared research activities in the fisheries in Burkina Faso. It also shows that merely trying to be interdisciplinary and transdisciplinary in a European-African research collaboration does not suffice, as hegemonies from male-dominated natural science and hegemonic postcolonial biases prevail. The case shows that such problems can remain unrecognized and that conflicts are perhaps needed to draw attention to these issues. The paper also addresses other views, which appear in the discussion of differences regarding gender, culture and multiple languages involved. The paper shows that the richest experiences and learnings stem from joint workshops and storytelling approaches, along with long-term interactions in the field. However, these require time and intensive interactions diverse participating actors in international, interdisciplinary and transdisciplinary project collaborations.

1.7. References

Bennett, J. W. (1996). Applied and action anthropology: Ideological and conceptual aspects. Current Anthropology, 37(1), 23–53.

Berkes, F. (1999). Sacred ecology. Traditional ecological knowledge, and resource management. Philadelphia: Taylor and Francis.

Berkes, F. and Folke, C. (eds.). (2000). Linking social and ecological systems. Cambridge, UK: Cambridge University Press.

Bornemann, B., Bernasconi, A., Ejderyan, O., Schmid, F., Wäger, P., Zingerli, C. (2017). Research on natural resources: The quest for integration revisited. GAIA, 26(1), 16-21.

Brockington, D., Duffy, R. and Igoe, J. (2008). Nature unbound: The past, present and future of protected areas. London: Earthscan.

Chabwela, H. and Haller, T. (2010). Governance issues, potentials and failures of participatory collective action in the Kafue Flats, Zambia. International Journal of the Commons, 4(2), 621–642.

Descola, P. (2013). Beyond nature and culture. Chicago: University of Chicago Press.

Ellen, R. (1982). Environment, subsistence and system. The ecology of small-scale social formations. Cambridge, UK: Cambridge University Press.

Ensminger, J. (1992). Making a market. The institutional transformation of an African society. Cambridge UK: Cambridge University Press.

Fairhead, J. and Leach, M. (1996). Misreading the African landscape: Society and ecology in a forest-savanna mosaic. Cambridge, UK: Cambridge University Press.

Faye, P., Haller, T. and Ribot, J. (2018). Shaping rules and practice for more justice. Local conventions and local resistance in Eastern Senegal. Human Ecology, 46(1), 15–25.

Foley, D. E. (1999). The Fox project: A reappraisal. Current Anthropology, 40(2), 171–192.

Galvin, M. and Haller T. (eds). (2008). People, Protected Areas and Global Change. Participatory Conservation in Latin America, Africa, Asia and Europe. Perspectives of the Swiss National Centre of Competence in Research (NCCR) North-South 3. Bern: Geographica Bernensia.

Haller, T. (2007). Is there a culture of sustainability? What social and cultural anthropology has to offer 15 years after Rio. In: Burger, P. and Kaufmann-Hayoz, R. (eds.). 15 Jahre nach Rio – Der Nachhaltigkeitsdiskurs in den Geistes- und Sozialwissenschaften: Perspektiven – Leistungen – Defizite (pp. 329–356). Bern: Schweizerische Akademie der Geistes und Sozialwissenschaften.

Haller, T. (ed.). (2010). Disputing the floodplains: Institutional change and the politics of resource management in African wetlands. African Social Studies Series 22. Leiden: Brill.

Haller, T. (2013). The contested floodplain: Institutional change of the commons in the Kafue Flats, Zambia. Lanham: Lexington.

Haller, T. (2016). Managing the commons with floods: The role of institutions and power relations for water governance and food resilience in African floodplains. In: Ostegard, T. (ed.). Water and food: Africa in a global context (pp. 369–397). London: The Nordic African Institute.

Haller, T. (2019). Towards a new institutional political ecology: how to marry external effects, institutional change and the role of power and ideology in commons studies. In: Haller, T., Breu, T., de Moor, T., Rohr, C., Znoj, H. (eds.). The Commons in a Glocal World: Global Connections and Local Responses (pp. 90-120). London: Routledge.

Haller, T. (2019). The different meanings of land in the age of neoliberalism: Theoretical reflections on commons and resilience grabbing from a social anthropological perspective. Land, 8(7), 104.

Haller, T., Acciaioli, G. and Rist, S. (2016). Constitutionality: Conditions for crafting local ownership of institution – Building processes. Society and Natural Resources, 29(1), 68–87.

Haller, T., Belsky, J. M. and Rist, S. (2018). The constitutionality approach: Conditions, opportunities, and challenges for bottom-up institution building. Human Ecology, 46(1), 1–2.

Haller, T., Fokou, G., Mbeyale, G. and Meroka, P. (2013). How fit turns into misfit and back: Institutional transformations of pastoral commons in African floodplains. Ecology and Society, 18(1), 34.

Haller, T. and Merten, S. (2008). "We are Zambians – don't tell us how to fish!" Institutional change, power relations and conflicts in the Kafue Flats fisheries in Zambia. Human Ecology, 36(5), 699–715.

Haller, T. and Merten, S. (2018). Crafting our own rules: Constitutionality as a bottom-up approach for the development of by-laws in Zambia. Human Ecology, 46(1), 3–13.

Haller, T. and Galvin, M. (2011). Challenges for participatory conservation in times of global change: Lessons from a comparative analysis and new developments. In: Wiesmann, U. and Hurni, H. (eds.). Research for Sustainable Development: Foundations, Experiences, and Perspectives (pp. 467–503). Perspectives of the Swiss National Centre of Competence in Research (NCCR) North-South 6. Bern: Geographica Bernensia.

Hardin, G. (1968). The tragedy of the commons. Science, 162(3859), 1243–1248.

Homewood, K. (2008). Ecology of African pastoralist societies. Oxford: James Currey.

Humboldt, A. (1852). Personal narratives of travels to the equinoctial regions of America. London: Henry G. Bohn.

Humboldt, A. (1853). The isthmus of Darien. Letter from Baron Alexander von Humboldt to Lionel Gisborne, Esq., M. A., C. E. Translated from the French original by The New York Times, NY.

Ingold, T. (2000). The Perception of the Environment. Essays on livelihood, dwelling and skill. London, New York: Routledge.

Lubrich, O. (ed.). (2009). Über die Urvölker von Amerika und die Denkmähler, welche von ihnen übrig geblieben sind: anthropologische und ethnographische Schriften. Fundstücke 27. Hannover: Wehrhahn.

McCabe, T. (1990). Turkana pastoralism. A case against the tragedy of the commons. Human Ecology, 18(1), 81–103.

Netting, R. (1993). Smallholders, householders: Farm families and the ecology of intensive, sustainable agriculture. Stanford, CA: Stanford University Press.

Neumann, R. P. (1998). Imposing wilderness. Struggles over livelihood and nature preservation in Africa. Berkley, CA: University of California Press.

North, D. (1990). Institutions, institutional change and economic performance. Cambridge, UK: Cambridge University Press.

Richards, P. (1985). Indigenous agricultural revolution. Ecology and food production in Africa. London: Hutchinson.

Robbins, P. (2004). Political Ecology: A Critical Introduction. Blackwell, Malden.

Rylko-Bauer, B., Singer, M. and Van Willigen, J. (2006). Reclaiming applied anthropology: Its past, present, and future. American Anthropologist, 108(1), 178–190.

Sahlins, M. (1972). The original affluent society, in Stone Age economics. Chicago, IL: Aldine.

Scott, J. C. (1998). Seeing like a state: How certain schemes to improve the human condition have failed. New Haven: Yale University Press.

Winterhalder, B. and Smith, E. A. (1985). Hunter-gatherer foraging strategies. Ethnographic and archaeological analysis. Chicago: University of Chicago Press.

Wulf, A. (2016). Alexander von Humboldt und die Erfindung der Natur. Munich: Random House GmbH.

Zingerli, C. (2010). A sociology of international research partnerships for sustainable development. European Journal of Development Research, 22(2), 217-233.

2. Soil classifications
Between material facts and socio-ecological narratives

Brice Prudat, Lena Bloemertz, Olivier Graefe, Nikolaus Kuhn

2.1. Introduction

Local environmental knowledge and enhanced community participation in research and implementation have been used for a better understanding of the lack of new technology implementation in local communities - for example, for water management or agriculture. The benefits of local knowledge can be tapped by including farmers and their understandings of local needs in economic development. Local soil knowledge is "the knowledge of soil properties (...) possessed by people living in a particular environment for some period of time" (Winklerprins 1999: 151). This knowledge integrates various environmental (e.g. soil, climate) and social criteria (e.g. techniques or labour force availability) that influence soil productivity. Local soil typologies are the way soils are named and grouped, and they aim to describe the local environment in order to support people to fulfil local necessities, such as for steady food production under high rainfall variability (Barrera-Bassols et al. 2006). Integrating local soil typologies and technical knowledge - for example, provided by international soil classifications - as much as possible without losing the essence of one or the other aims to make local environmental knowledge accessible to outsiders. The integration of local and technical knowledge intends finally to improve agricultural management practices (Winklerprins 1999). The comparison and combination of these typologies require building an important body of knowledge that includes farmers' and scientific soil knowledge. However, the translation, simplification and codification of this new mixed knowledge is often found to be the cause of the failure to use local knowledge for development (Briggs 2013, Pottier et al. 2003). Although these processes are necessary to transfer explicit information to a wider audience (externalization), more sharing and a participatory research agenda are helpful.

The issues raised in using this mixed knowledge are that local soil knowledge is fragmentary and dynamic, and therefore difficult to translate into or combine with more centralized, international and static knowledge systems (Agrawal 1995, Rathwell et al. 2015, Sillitoe, 2010). Local soil types are not strictly defined and are described on a comparative basis (Sillitoe 1998), and can therefore vary from one village to another (Barrera-Bassols and Zinck 2003) depending on the socio-cultural context (Sillitoe 1998) and the local environmental conditions (Niemeijer and Mazzucato 2003). This flexibility and dynamism pose fundamental challenges to scientific endeavours focusing on the (more) strict, systematic and context-independent classification of objects (Ellen et al. 2000, Hobart 2002, Pottier et al. 2003).

In this article, we will spell out what it means to do environmental research in the interplay of local, implicit soil knowledge and international, explicit soil classification. We will contextualize the Oshikwanyama soil typology from north-central Namibia and its relation to international classification, and discuss the advantages of local knowledge for soil quality assessment. Thereafter we will reflect on issues regarding the collection, translation and selection of local soil knowledge. As a part of this, we will reflect on the experience of participatory approaches from the perspective of the first author, trained as a natural scientist.

2.1.1. Ohangwena region and villages

Ohangwena region in north-central Namibia is characterized by the endorheic Cuvelai drainage basin in the westernmost part and the Kalahari Sandveld in the central and eastern part (Figure 1). The climate is semi-arid and subtropical, with large inter- and intra-annual variability (Mendelsohn et al. 2000). Oshikwanyama-speaking communities immigrated into today's Ohangwena region during the late nineteenth century and moved eastwards during the first decades of the twentieth century (Kreike 2004). The Cuvelai drainage basin has to a large extent been converted into crop fields for small-scale (1-4 ha) non-commercial agriculture of rain-fed pearl millet (*Pennisetum glaucum*; Mendelsohn et al. 2000).

For our study, we selected three village areas in the western Ohangwena region, based on dialect homogeneity (Oshikwanyama) and environmental

Figure 1: Overview of Africa, Namibia and north-central Namibia

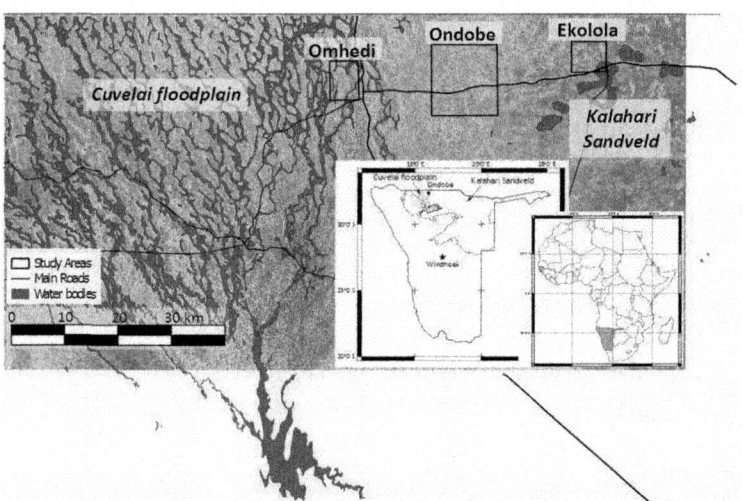

Maps of Africa, Namibia and a satellite image (maps.google.com, retrieved in July 2016) of north-central Namibia with the Cuvelai floodplain (west), the Kalahari Sandveld (east) and location of the three study areas (squares). Water channels (*iishana*) and temporary ponds are in blue, vegetation and bare soil appear in green and in orange respectively.

heterogeneity, including vegetation and soils (Figure 1). These villages (Omhedi, Ondobe and Ekolola) are on a west-east gradient, representing edaphic and vegetation differences with decreasing influence of the Cuvelai River eastwards. Omhedi, the westernmost area, is situated in the Cuvelai drainage basin with active ephemeral water streams (*iishana*). Ondobe is located between the drainage basin (mostly inactive *iishana*) and the Kalahari Sandveld that lies east of Ondobe. Ekolola is characterized by the Kalahari Sandveld and is largely covered by deep loose sand deposits, forests and extended temporary pans (Mendelsohn et al. 2000).

2.1.2. Collecting local soil knowledge

The data collection for our study was carried out during various extended field stays between February 2013 and June 2014. We used semi-structured interviews to construct a local soil typology and to understand local farmers'[1] soil quality perception. Most interviews were conducted in the farm homestead, promoting abstract discussion about soil types and definitions. This approach was chosen because it helped the researchers to create a local soil typology that can be extrapolated to a regional scale. However, some interviews were also conducted during transect walks through the field or in front of soil pits, both leading to discussions concerning micro-level soil transitions relevant for management practices, as suggested by Oudwater and Martin (2003).

In total, we conducted 87 interviews on 46 farms, mainly in Ondobe (50 interviews / 21 farms). From March to June 2014, we collected additional interviews from Omhedi (19 interviews / 15 farms) and Ekolola (18 interviews / 10 farms). In most cases, the head of the household (mostly men above the age of 60) was interviewed, as the family members within the households suggested this. Mostly, the interview language was Oshikwanyama and translation into English was provided by a translator from Ongwediva (Oshana region) without a background in soil science. This young woman was the translator for the entire duration of the data collection period. The continuous collaboration enabled the research team to build a common knowledge and language. The quality of the information collected improved during the period of study, given that both the researchers' knowledge concerning local soils and the interpreter's skills considerably increased (similar observation was done by, for example, Oudwater and Martin 2003).

All the interviews were audio-recorded and the English oral translation was transcribed. The most relevant parts of the interviews were transcribed in Oshikwanyama and translated into English. We used MAXQDA 11 (VERBI GmbH, 2014) to organize and classify the interviews.

2.1.3. Scientific soil description

We scientifically described 28 soil profiles in cultivated fields in Ondobe (21), Omhedi (3) and Ekolola (4). These profiles were classified by the farmers as *omutunda* (14), *ehenge* (4), *omufitu* (4), *elondo* (3) and *ehenene* (3). We selected more *omutunda* for the analysis given the high agricultural value of this local soil type and its prevalence in the cultivated area.

---

1   All informants involved in this study are called farmers, despite the fact that crop cultivation is not necessarily their main economic activity.

We classified the described soils using two scientific soil classifications: the World Reference Base for Soil Resources 2014 (WRB; IUSS Working Group WRB 2014) and the Fertility Capability Soil Classification (FCC; Sanchez et al. 2003). Both require the analysis of various chemical and physical properties and exclude properties that reflect short-term changes. The WRB aims at identifying pedological structures and uses properties that are mostly the outcome of long-term soil evolution (aside from anthropogenic soil modifications). On the other hand, the FCC aims at highlighting limiting factors for crop production, specifically for tropical soils, and deals with properties "that are either dynamic at time scales of years or decades with management, as well as inherent ones that do not change in less than a century" (Sanchez et al. 2003: 157).

Both classifications present a hierarchical classification structure. The WRB's Reference Soil Groups are "differentiated mainly according to primary pedogenetic process[es]" (IUSS Working Group WRB 2014: 5) and the FCC's substrate reflects the soil type (texture).

2.1.4. The Oshikwanyama soil units

The body of mixed local-technical knowledge summarized in Table 1 is the result of previous studies (Hillyer et al. 2006, Newsham and Thomas 2011, Rigourd and Sappe 1999, Verlinden and Dayot 2005) and the interviews conducted during the current study. Interviewees described the local soil units mostly based on soil consistency (hard or soft) and colour shade (dark or light), as well as the sensitivity to waterlogging conditions.[2] These properties are related to soil suitability for cultivation - for instance, workability and fertility. Soil hydrology has a strong influence on agricultural suitability and therefore on local soil typology and, given the rainfall irregularity, both waterlogging and soil drought occur frequently during the rainy season. The five soil units described in Table 1 can be used as cornerstones for soil quality evaluation as they represent important soil processes and characteristics for crop cultivation (waterlogging risks, texture).

---

2   Waterlogging conditions indicate soil saturation with water and strongly inhibit roots' respiration.

Table 1: List of local soil types

	Soil type attribute		
	Water related characteristics	Consistence	Colour shade
Omu-tunda	No waterlogging High water retention capacity Dries out quickly	Hard	Dark/black
Omu-fitu	No waterlogging Low water retention capacity	Loose	Dark or light
Elondo	No waterlogging	Intermediate	Intermediate
Ehenge	Waterlogging risk Dries out very slowly	Loose	Light/white
Ehenene	Waterlogging risk Low water retention capacity Dries out quickly	Hard	Light/white

2.1.5. Local soil types compared to international classifications

According to the WRB, the 28 soil profiles would be classified as arenosols (17), regosols (10) or calcisols (1) (Figure 2, left). These reference soil groups are almost exclusively determined by soil texture: sandy and loamy sand soils are classified as arenosols, while sandy loam and finer soils are regosols. The fertile soil (*omutunda*) has a loamy sand or sandy loam texture (<90% sand), while the other soil types have a sand texture (>90% sand), excluding *ehenene*.[3] This textural difference is significant for productivity and soil management strategies in north-central Namibia. Both the WRB and FCC classifications group all sand and loamy sand soil types into a single reference soil group (arenosols in WRB) or substrate (S in FCC) and describe these classes as having low chemical fertility. Farmers in the western Ohangwena region contradict this evaluation and consider loamy sand soils as good soils to grow pearl millet in north-central Namibia (i.e. *omutunda*). A better differentiation

---

3   Productivity in *ehenene* is limited by poor water infiltration, high pH and sodic conditions.

of the soils depends on the identification of other characteristics (e.g. base saturation, colour). Given the gently undulating landscape of the region, the spatial pattern of the distribution of local soil types is generally related to micro-topography (elevation differences of a maximum of a few metres), which results from the variable intensity of influence of the Cuvelai River. In contrast, the WRB classification is more driven by macro-topography. The lack of macro-topography in northern Namibia renders the linkage of local soil types' distribution to landforms (slope, plateau), and therefore to WRB classes, difficult.

Figure 2 gives an overview of the relation of the local soil typologies to the two scientific classification systems. This schematic representation highlights that soils are divided into different classification groups (e.g. distribution of *omutunda* symbols between regosols, arenosols and calcisols, or S, SL and L). It becomes clear that a simple translation of local soil types into WRB or FCC soil classes is not straightforward.

Figure 2: Schematic representation of the described soil profiles classified in WRB and FCC classifications

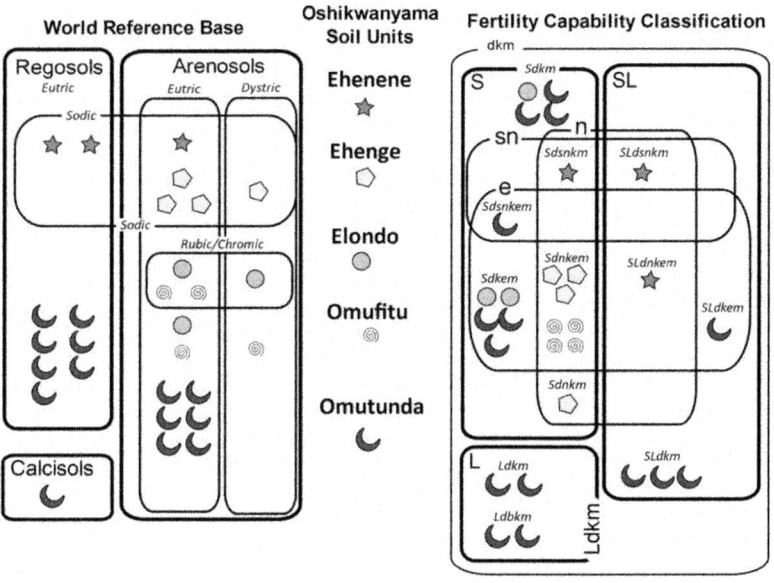

Soil hydrology has a great influence on soil productivity, mostly in relation to rainfall variability. Indeed, both waterlogging and soil drought conditions, even for short periods of time (a few weeks) during the growing season, strongly reduce yields. The occurrence of these short events is not taken into consideration in either the WRB or the FCC.

Based on our findings, we argue that the translation of local soil types into international classifications is not relevant to evaluate soil quality in north-central Namibia. Indeed, the two selected international classifications use soil texture and soil hydraulic conductivity; both treat properties differently than farmers would, yet these properties are highly relevant in the local context and strongly influence soil productivity. Despite a seeming misfit, the local knowledge and the international soil classification complement each other. However, they are more likely to be of interest to researchers and experts than to the farmers themselves.

2.1.6. Advantages of combining local and scientific knowledges

Including farmers' knowledge and trying out a shared research arrangement helped us to highlight important limiting factors for agricultural production in the local context (e.g. soil water availability during the early growing stage) and soil characteristics that are difficult to detect during conventional soil surveys (waterlogging conditions, micro-scale soil heterogeneity). For these reasons, we support the use of local assessment as an entry point to understand and assess soil quality at the regional level. We suggest that stepping back from numerical and quantitative data - without excluding them - can improve soil fertility assessments on local scales. In comparison with natural science surveys, local soil knowledge presents many advantages. Firstly, it assesses quality based on needs; secondly, it requires no laboratory analysis; thirdly, it includes the most relevant characteristics locally - for example, soil water characteristics and humidity variability; and fourthly, it helps to reduce the number of variables that must be assessed for a locally relevant soil quality evaluation (in our case, for pearl millet cultivation) in a specific climate. Modern methods can, however, adequately complement the local soil knowledge by providing standardization (Niemeijer and Mazzucato 2003) and tools for extra-regional communication. Therefore, the two soil knowledge systems should be used in a complementary way.

2.2. Issues regarding the participatory approach in natural sciences

By increasing our understanding of local soil knowledge we faced important difficulties. We supposed that similar issues came up in other studies that aimed to understand local soil knowledge and therefore reflected on our experiences against the literature. We tended to conclude that despite the usefulness of local soil typology for soil quality assessments, the use of local knowledge has limitations in the communication, which will be explained below using direct quotes from our interviews.

2.2.1. Translations of the concept of "soil"

For soil scientists, soils are vertical successions of horizons, which is a restricted concept that is not recognized in many cultures (Barrera-Bassols et al. 2009). The Oshikwanyama word *edu* [translation of the word "soil"] integrates broad concepts related to space, from landscape to sand. The range of meanings of *edu* led to misunderstandings and confusions about the spatial scale during many interviews. In particular, *omutunda* and *omufitu* were used to point to general landscapes as well as to specific soil types:

> "All over here, people are in *omufitu* area. The other side of the village is in *omutunda*. Mostly you find *ehenge* in *omutunda* area (LM, 80, Oipapakane).[4]
> But at *omufitu* is where you find *ehenge* (AA, 70, Oipapakane).
> All the parcel, like Martha, Kelly, Kalola, here, we are in *omutunda* area, but there are different soil types, like small *omahenene*, or *ehenge*. But in this *omutunda* area, you cannot find *efululu*"[5] (LS, 65, Ondobe).

These quotes show that there is a need to contextualize local knowledge and to acknowledge variation within specific soil types relevant for local users. For example, large areas like villages are referred to as *omutunda* (Ondobe village) or *omufitu* (Ekolola village), thereby considering and understanding these units as landscapes features. Within this context, *omufitu* was described as an "area that is not cleared [of trees] even a little" (KS, 65, Ohengobe). However, shortly after, when the focus shifted to soils, the same informant said: "If I

---

4 To keep the informants anonymous, we used a code that indicates: 1) a two-letter name; 2) the farmers' age; and 3) the study area of the farm.
5 *Efululu* mostly refers to a type of fine loose sand.

cultivate on *omufitu*, the *edu* will look red", clearly referring to a soil. *Edu* also refers to soil layers and sand. In the expression, *edu li hapu*, which literally means "a lot of sand", *edu* takes the meaning of sand.

These examples illustrate the permanent differentiation between scales during discussions, while the scales of *edu* are integrated into one another (e.g. *omutunda* soil type in an *omufitu* landscape). Other authors point out the existence of a similar word, with a large range of meaning, in different African regions (Birmingham 2003, Lamers and Feil 1995, Niemeijer and Mazzucato 2003, Osbahr and Allan 2003). "Soil" in English or "terre" in French are two examples from Europe that also cover very large concepts related to space. These overlapping definitions increase the risk of confusion and misunderstandings. However, as used by Hillyer et al. (2006), these soil/landscape names have the advantage of including uncultivated lands, i.e. not ploughed.

From our observations, informants talked more about the landscape definition of *edu* when referring to abstract spatial concepts (e.g. villages) and focused on agricultural soil when referring to their fields. As an example, CH (65, Efidi) claimed that all his farm was in *omufitu*, but during transect walks it was possible to find *omutunda*, *ehenene* and most of the field was *ehenge*, while no *omufitu* was described.

The potential confusion regarding the concept of soil was mostly resolved by specifically improving the precision of communication between the translator and the researcher. Parts of the interviews were transcribed together to differentiate between these scales.

2.2.2. Intergrades

Local typology is highly adaptive to conditions and adjusts to changes - and even the interview context - by the extended use of combined soil names ("intergrades", after Krogh and Paarup-Laursen 1997). The intergrades are good indicators for land degradation or improvement in relation to, for example, land management techniques, and are a way to emphasize certain properties (Birmingham 2003). A soil might, for example, be related to *omutunda* to emphasize its good productivity but *omufitu* to emphasize its loose consistency in comparison to other locations.

> "This soil is a mixed soil of *ehenge* and *ehenene*, it's like a *ehenene-henge* (CH, 65, Efidi).

... because all where is *omutunda-henene* if it rains you can sink ... (AA, 75, Oipapakane).
Close to the fig tree and there, it is *ehenge*. In between, there is *omufitu* and it is *omufituhenge*" (KF, 65, Etomba).

During the first interviews at each farm, held in the house, these intergrades appeared only rarely. During transect walks, while trying to understand more details about local soil types, they appeared more frequently. Following the insistence of the interviewers to find the "real" *omutunda*, or the "real" *omufitu*, these intergrades appeared in high proportion.

2.2.3. Local experts

It proved to be very difficult to collect information concerning local soil types in Oshikwanyama from literate or schooled people. English speaking informants (local elite, ministry officer) explained the soil diversity in terms of three "soil" types, namely sandy soils, loamy soils and clay soils. We assume that these informants are aware of the soil diversity and local soil types, but it seemed clear that they considered the information learned in school as more valuable or more in line with the expectations of an outsider in comparison to the local knowledge.

2.2.4. Accuracy of descriptions

The accuracy of soil descriptions collected from the interviews depends on the local soil unit considered. This is related to the various values given to the soil for a specific unit. Verlinden and Dayot (2005) observed that depending on the indigenous land units, soil characteristics have various levels of importance. This explains that soils are more narrowly defined in soil units when used for cultivation.

2.2.4.1. *Omufitu* and the importance of soil versus vegetation information

Omufitu largely refers to areas where bushes are still present, and despite awareness of the scale issue (soil/landscape) of *omufitu*, it was difficult to get clear descriptions for *omufitu* as a soil type (e.g. soil colour; Table 1), mostly because soil characteristics are less important than vegetation in defining this unit.

> "*Omufitu*, they are different. I think that an area is called *omufitu* if the area is not cleared from trees" (KS, 60, Ohengobe).
>
> "You can find *omufitu* that will give you good food [...], but some of your neighbours with *omufitu* might not get anything from it" (NW, 70, Etomba).

In *omufitu*, soils did not matter much in a relatively recent past, as they were mostly kept for firewood and grazing. Therefore, soil characteristics are not important and soils have different qualities. However, *omufitu* is today increasingly cleared and cultivated, and the number of distinguished *omufitu* soil types may therefore grow with the rising cultivation rate (for a similar example in eastern Burkina Faso, see Niemeijer and Mazzucato 2003).

2.2.4.2. *Omutunda* and the relativity of soil quality in relation to the surrounding environment

In contrast to *omufitu*, soil characteristics are well defined for *omutunda* in the literature and during the interviews, because these soils have been largely turned into fields. This accuracy does not mean that all *omutunda* are similar, but they share a set of characteristics. Almost all informants described them as hard and dark soils.

The quality of a soil in a field is often compared with the other soils in one's own fields, however, and soils are defined comparatively, as observed by Birmingham (2003). Furthermore, it has been observed in various studies that soil quality ascriptions of local soil types may vary depending on various parameters such as individual perceptions (Barrera-Bassols et al. 2006), intended uses (e.g. agricultural versus housing; Gray and Morant 2003, Niemeijer and Mazzucato 2003) or specific environmental conditions in the surroundings (Gray and Morant 2003).

Omutunda was defined as the best soil for pearl millet by most informants.

> "There are different types of *omutunda*. At Tate S., *omutunda* is not good because there is stone; it will only be good soil when you add cow dung" (LN, 65, Omhedi).

Omutunda is "where you feed" (CK, 65, Ohandiba), and farmers tend to describe the most productive part of their land as *omutunda*. The *omutunda* described in or close to the Cuvelai drainage system (Omhedi and Ondobe) is finer and darker than the *omutunda* found in the Kalahari woodland biome (Ekolola). We can show this difference using technical parameters (pH, fine particles content, colour shade; Figure 3). This result indicates that the productivity potential of *omutunda* is lower in the woodland biome than in the floodplain environment, which was acknowledged by the farmers themselves:

> "The soil [*omutunda*] ... inside the country [floodplain] breastfeeds on *iishana* ... it is hard not like ours [Ekolola area]" (TN, 70, Ohandiba).

A farmer from Omhedi (eastern floodplain) used intergrades to illustrate similar soil quality difference:

> "... that is *omufitu-tunda* ...Yes, because did you say it is at Eengonyo [Ekolola area]? *Omufitu-tunda* because that *omutunda* does not occupy a big area" (IS, 75, Omhedi).

In this example, the region (Eengonyo village) is important information to claim that the *omutunda* described is an intergrade between *omutunda* and *omufitu* (*omufitu-tunda*), because *omutunda* "does not occupy a big area". Indeed, Eengonyo (in the Ekolola area) is situated in an area with a lot of deep sands and forest (*omufitu*).

2.2.4.3. Management and rainfall as influencing soil quality

Another aspect that needs to be taken into consideration is that actual soil quality (the yield achieved on a specific soil in a specific year) is strongly influenced by inputs (e.g. fertilizer, labour, rainfall). Even sandy soils can be productive if fertilizers (mostly manure) are used appropriately.

> "*Omutunda* and *omufitu* produce the same, but it strongly depends on manure availability" (LS, 65, Ondobe).

Consequently, the ongoing decline of livestock density in villages and/or the use of tractor ploughing is leading to a soil fertility decline in many areas. Land degradation was indicated by the transformation of *omutunda* soils (mostly) into other local soil types, which are, by definition, less fertile.

> "Originally our soil was *omutunda*, but then it became *ehenene*..." (CP, 40, Oilyateko).

Conversely, old homestead or kraal locations turn local soil into *omutunda*:

> "The field was just *ehenge*, but it has changed and it looks like *omutunda*, because when we shift the house or the kraal, or apply manure, it changes *ehenge* into something else" (VH, 45, Oilyateko).

Figure 3: Regional variability of omutunda

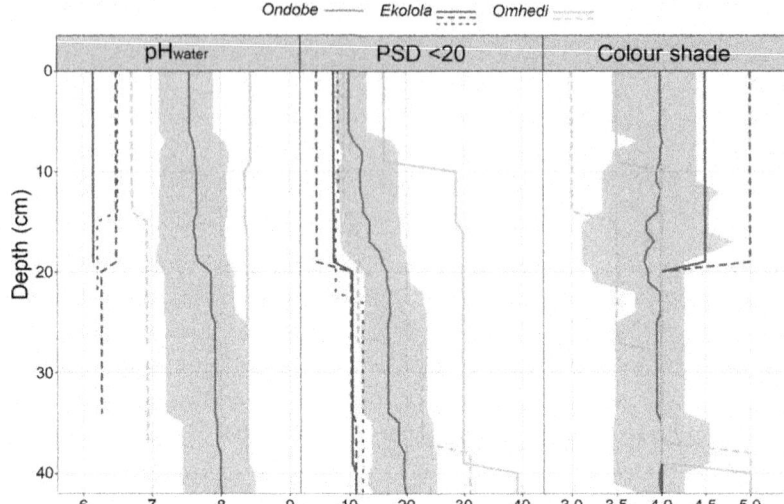

The central line represents the median value of described omutunda in Ondobe, the shaded area represent 25 and 75 percentiles. Light grey lines represent the individual profiles of omutunda described in Omhedi and dark grey lines represent the individual profiles of omutunda described in Ekolola. The results indicate that omutunda from Ekolola has a lower pH, finer particle content (PSD <20), and is lighter (higher colour shade).

> "If you go and look at my crops there you will see how the cattle changed the soil. Now sand soil became *omutunda*" (MJ, 60, Omhedi).

Therefore, as soil quality can be altered rather fast through the application of manure or through over-exploitation, correspondingly the soil type can also change in a relatively short period. This highlights that labelling works with culturally built landscapes.

> "The availability of water in soil is the most limiting factor for agricultural productivity in north-central Namibia. Each soil type has a different productivity potential depending on rainfall scenarios (intensity, amount and distribution). Most obvious examples in north-central Namibia are the productivity of *omutunda* and *ehenge* [Table 1]. *Omutunda* is productive when rainfalls are frequent but has a poor yield if extended dry spells occur. On the other hand, *ehenge* is more productive during rainy seasons with below average

rains. During these years, it will not experience waterlogging conditions. [...] if it rains a lot, *omutunda* does not grow good millet. If there are short rains, then we will harvest at *ehenge*" (JP, 60, Omhedi B).

Consequently, *ehenge* is mainly cultivated to minimize the risk of crop failure during years of poor rainfall. In general, farmers will always plant on a mixture of different soils, not to aim for the best possible yield, but to reduce risks, as described in other studies (Briggs and Moyo 2012, Gray and Morant 2003, Krogh and Paarup-Laursen 1997, Niemeijer and Mazzucato 2003, Osbahr and Allan 2003). This behaviour is described by game theory as a minimax strategy (e.g. Lipton 1982).

2.3. Participatory research in natural sciences: reflections and challenges

2.3.1. Expectations and managing data

During this study, a young natural scientist collected local soil typologies and compared them to international classifications. Many farmers had some difficulties valuing their knowledge with regard to the white male scientist's knowledge. However, once they understood the goals of the study, they were very keen and proud to share their knowledge. Before starting the data collection, both interviewers (scientist and translator) had some expectations regarding the information that would and should be collected. Partly for that reason, early during the data collection period, we concluded that the main local soil types (*omutunda*, *omufitu*, *ehenge* and *ehenene*) can be defined using a limited amount of information. The information summarised in Table 1 is in line with the data collected during each interview, with few exceptions.

However, during the 87 interviews, a large quantity of information concerning soils was collected which did not correspond directly with our expectations. Thus, a continuous selection of information was necessary to do the analysis and to establish Table 1. The information that seemed to the scientist and the translator to be the most relevant information was selected, transcribed, codified and used for further analysis (comparing scientific and local soil knowledge, translation and classification). This led to a potentially biased or incomplete soil typology comparison, as the information which remained unclear, inconsistent or confusing was dismissed. The dismissed information might, however, have held important implicit local knowledge. For example, *ehenge* and *ehenene* are two different soil types described by most farmers and in the literature (Table 1), but two informants suggested that these soils are the same:

> "*Ehenge* is *ehenene*. Normally it is called *ehenge*, but it is *ehenene*. *Ehenge* is also *ehenene*, or vice-versa" (JD, 55, Omhedi).

Dismissed because it does not correspond to most opinions, this information may indicate that these two soils are very closely related to each other and differentiating them is irrelevant for some informants. This information highlighted the connection between these soils in relation to waterlogging probability and the presence of a hardened layer at various depths.

Another piece of evidence dismissed is that *omufitu* is considered either good or bad during droughts. Given the high permeability of this soil, we would favour the information that stipulates that *omufitu* is bad during droughts because it does not hold water. This was also the most commonly found information. However, we should not exclude that some *omufitu* might be good (or better) during droughts:

> "The entire field is the same soil [*omufitu*]. It is not suitable for millet ... *Omufitu* does not lose water moisture underground. So, with *omufitu*, people will have at least some millet this year [2013, drought]" (MH, 70, Ondobe).

2.3.2. Dealing with complexity

This information exclusion process was required to communicate and reduce the complexity of local knowledge. The information collected during the interviews can be divided into two levels of local knowledge: i) the local soil groups, with clearly defined characteristics; and ii) the information regarding specific locations, specific terms and incoherence highlighting soil processes. The information found on the first level can be transferred to and used by outsiders without many difficulties. It includes the most explicit information regarding land management (e.g. hardness, waterlogging risks). However, it should not be forgotten that summaries and externalization give only a glimpse into local knowledge and therefore need to be used carefully. The implicit information found in the second level of knowledge is more difficult to access and to make explicit to outsiders, as it varies from person to person. The set of information held in this body of knowledge cannot be codified, classified or generalized, and therefore needs to be collected directly (personalized) without any (or only limited) translation or intermediaries. Understanding all variations of this knowledge renders it difficult for outsiders to use. Through externalization to outsiders, a large proportion of this knowledge is lost and misinterpreted (or over-interpreted). It loses accuracy, but could be used in combination with other knowledge.

The summary table (Table 1) is the result of parallel processes of collection and selection performed by the translator and the authors. Identification of the key properties enabled them to use and communicate a simple soil typology that can thereafter be used, for example, in soil quality assessment (Prudat et al. 2018). This simplification often occurs when scientists categorize and communicate environmental and soil local knowledge to outsiders (Barrera-Bassols et al. 2006). However, as discussed above, the soil quality of local soil types is actually more complex than its presentation to readers.

Oudwater and Martin (2003) emphasize that social scientists do not have the necessary tools to understand soil typologies. Conversely, collecting farmers' soil knowledge is not as simple as visiting various farms and asking questions. The understanding of a local soil typology implies that soil scientists experience the local contexts for extended periods, but also to get acquainted with semi-structured interview methods and qualitative data analysis.

2.4. Conclusion and perspectives

The type of shared research we engaged in was to be open to local labelling of soils. We realized through this process that soil classes refer not only to general soil properties, but also to how soils are contextualized regarding the aim of production and the view on environmental factors, and they are set in relation to other areas observed in the neighbourhood.

We discussed the benefits and limitations of using the local soil typology to evaluate soil quality in regions with poorly developed and sandy soils. To collect farmers' knowledge about soils and evaluate their perception of soil quality, soil scientists need to engage with the local community. The explicit knowledge that can be collected and transferred (externalized) is a generalized and codified knowledge, while many implicit details mentioned by a few are excluded. The exclusion of a large proportion of the knowledge collected is criticized, but this process is necessary in order to remain usable to outsiders. It should be emphasized when externalizing this knowledge, however, that it does not represent the opinion of all members of the community and that the "local" knowledge has been de-localized and large proportions of implicit knowledge excluded. The possibility of collecting and understanding such knowledge during rapid appraisal should therefore be looked at critically, because local knowledge (e.g. of soils), is based on a comparative basis and changes over time and space.

The high variability and confusing answers collected were frustrating. However, the value of the knowledge accumulated should not be underestimated. We show with our understanding that the classification of natural objects (e.g. soils) is the result of a codification process performed by "experts" (elders, academics), and aims at simplifying the diversity of objects (e.g. *omutunda* defines fertile areas). This classification will vary between experts, and depends on the context in which the object exists and is used.

In general, the main advantage of using a participatory approach in natural sciences is commonly thought to be the involvement of communities in the research. We would argue the other way around, and that it works both ways, in that a participatory approach, by forcing researchers to invest time in the community, engages the researchers in the social context in which the soils are used, giving a broader perspective than the soil itself. In this way, research moves from participatory (including the community in the research) to observatory research (including the researcher in the community). Engaging in local knowledge together with local actors allows us to come to a view on shared knowledge which cannot be reached by rapid methods, but by participating and observing. This emic position opens new perspectives for further researches and further discussions with scientists in various fields of study.

Acknowledgements

We would like to thank Florence Botin for her proofreading, Claudia Zingerli and Tobias Haller for their highly valuable comments, Vladimir Wingate, Romie Nghitevelekwa and Laura Weidmann for their assistance with fieldwork, Martha Fillemon for her invaluable translations and patience, and Alex Verlinden for his academic guidance through indigenous knowledge. The project was funded by the Swiss National Science Foundation and the German Research Foundation.

2.5. References

Agrawal, A. (1995). Dismantling the divide between indigenous and scientific knowledge. Development and Change, 26(3), 413–439.

Barrera-Bassols, N. and Zinck, J. A. (2003). Ethnopedology: A worldwide view on the soil knowledge of local people. Geoderma, 111(3), 171–195

Barrera-Bassols, N., Zinck, J. A. and van Ranst, E. (2006). Symbolism, knowledge and management of soil and land resources in indigenous communities: Ethnopedology at global, regional and local scales. CATENA, 65(2), 118–137.

Barrera-Bassols, N., Zinck, J. A. and van Ranst, E. (2009). Participatory soil survey: Experience in working with a Mesoamerican indigenous community. Soil Use and Management, 25(1), 43–56.

Birmingham, D. M. (2003). Local knowledge of soils: The case of contrast in Côte d'Ivoire. Geoderma, Ethnopedology, 111(3–4), 481–502.

Briggs, J. (2013). Indigenous knowledge: A false dawn for development theory and practice? Progress in Development Studies, 13(3), 231–243.

Briggs, J. and Moyo, B. (2012). The resilience of indigenous knowledge in small-scale African agriculture: Key drivers. Scottish Geographical Journal, 128(1), 64–80.

Ellen, R. F., Parkes, P. and Bicker, A. (eds.) (2000). Indigenous environmental knowledge and its transformations: Critical anthropological perspectives. Studies in Environmental Anthropology 5. Amsterdam: Harwood Academic.

Gray, L. C. and Morant, P. (2003). Reconciling indigenous knowledge with scientific assessment of soil fertility changes in southwestern Burkina Faso. Geoderma, Ethnopedology, 111(3–4), 425–437.

Hillyer, A. E. M., McDonagh, J. F. and Verlinden, A. (2006). Land-use and legumes in northern Namibia – The value of a local classification system. Agriculture, Ecosystems and Environment, 117(4), 251–265.

Hobart, M. (2002). An anthropological critique of development: The growth of ignorance. London: Routledge.

IUSS Working Group WRB. (2014). World reference base for soil resources 2014. International soil classification system for naming soils and creating legends for soil maps. World Soil Resources Reports. Rome: FAO.

Kreike, E. (2004). Re-creating Eden: Land use, environment, and society in southern Angola and northern Namibia. Portsmouth: Heinemann.

Krogh, L. and Paarup-Laursen, B. (1997). Indigenous soil knowledge among the Fulani of northern Burkina Faso: Linking soil science and anthropology in analysis of natural resource management. GeoJournal, 43(2), 189–197.

Lamers, J. P. A. and Feil, P. R. (1995). Farmers' knowledge and management of spatial soil and crop growth variability in Niger, West Africa. Netherlands Journal of Agricultural Science, 43(4), 375–389.

Lipton, M. (1982). Game against nature: Theories of peasant decision-making. In: Harriss, J. (ed.), Rural development (pp. 258–268). London: Routledge.

Mendelsohn, J. M., El Obeid, S. and Roberts, C. (2000). A profile of north-central Namibia. Windhoek: Gamsberg Macmillan Publishers.

Newsham, A. J. and Thomas, D. S. G. (2011). Knowing, farming and climate change adaptation in north-central Namibia. Global Environmental Change, 21(2), 761–770.

Niemeijer, D. and Mazzucato, V. (2003). Moving beyond indigenous soil taxonomies: Local theories of soils for sustainable development. Geoderma, Ethnopedology, 111(3–4), 403–424.

Osbahr, H. and Allan, C. (2003). Indigenous knowledge of soil fertility management in southwest Niger. Geoderma, Ethnopedology, 111(3–4), 457–479.

Oudwater, N. and Martin, A. (2003). Methods and issues in exploring local knowledge of soils. Geoderma, Ethnopedology, 111(3–4), 387–401.

Pottier, J., Sillitoe, P. and Bicker, A. (2003). Negotiating local knowledge: Identity and power in development. London: Pluto.

Prudat, B., Bloemertz, L. and Kuhn, N. J. (2018). Local soil quality assessment of north-central Namibia: Integrating farmers' and technical knowledge. SOIL 4, 47–62.

Rathwell, K., Armitage, D. and Berkes, F. (2015). Bridging knowledge systems to enhance governance of environmental commons: A typology of settings. International Journal of the Commons, 9(2), 851–880.

Rigourd, C. and Sappe, T. (1999). Investigating into soil fertility in the north central regions. In: Kaumbutho, P. G. and Simalenga, T. E. (eds.) Conservation tillage with animal traction. Harare: ATNESA.

Sanchez, P. A., Palm, C. A. and Buol, S. W. (2003). Fertility capability soil classification: A tool to help assess soil quality in the tropics. Geoderma, Ethnopedology, 114(3–4), 157–185.

Sillitoe, P. (1998). Knowing the land: Soil and land resource evaluation and indigenous knowledge. Soil Use and Management, 14(4), 188–193.

Sillitoe, P. (2010). Trust in development: Some implications of knowing in indigenous knowledge. Journal of the Royal Anthropological Institute, 16(1), 12–30.

Verlinden, A. and Dayot, B. (2005). A comparison between indigenous environmental knowledge and a conventional vegetation analysis in north central Namibia. Journal of Arid Environments, 62(1), 143–175.

Winklerprins, A. M. G. A. (1999). Insights and applications local soil knowledge: A tool for sustainable land management. Society and Natural Resources, 12(2), 151–61.

3. Action research and reverse thinking for anti-desertification methods
Applying local revegetation techniques based on the ecological knowledge of local farmers in the Sahel of West Africa

Shuichi Oyama

3.1. Introduction

"In November 2001, I observed the men of a Hausa village in Niger spending the afternoon listening to the radio and playing cards for money. When the sun set in the west, they would stop playing cards and return to their homes. They cleaned their livestock yards using rakes and collected the waste. The waste was then carried to their own farmland. The annual millet and cowpea crops had already been harvested and cows, goats, and sheep were left to graze the harvested fields, where they would eat the stems and leaves of the crops. One man put the waste onto an ox cart his 12-year-old boy was driving. Another man placed the waste into an 80 cm diameter iron pan and carried it on his head. The waste included compostable organic matter as well as not compostable plastic and metals. The heat of the sun was still strong in the evening and they were dripping with sweat. Women were also involved in carrying waste from the homesteads to the fields. They wrapped the waste in cloth and carried it on their heads."

This diary entry in the fieldnote-like description of a common scene in a small village in southern Niger stands at the beginning of a puzzle with which this paper deals. At first, I watched such situations without understanding that the waste, and non-compostable waste, was used by local people to fertilize and enrich their soils. This fact I came to understand only when I did participatory observation combined with experiments and helped them to carry the waste to the farmland, where it was placed on the ground. This experience in a Hausa village inspired my research activities and led to my quest for a better understanding of environmental restoration based on the ecological knowledge of Hausa farmers, who use non-organic waste to fertilize their land. How is this possible? How did it impact my collaborative research? And how does it challenge natural science discourse on land restoration in the Sahel by focusing on local ecological knowledge? These are questions explored in this article.

3.2. Desertification in the Sahel region

In the Sahel region, on the southern edge of the Sahara Desert, desertification is a serious problem. It occurs through a combination of natural (irregular rainfall, drought and poor soil fertility) and anthropogenic factors (over-cultivation, over-grazing and firewood collection). The rapid increase in the human population of the Sahel is also considered to be a fundamental high-pressure driving force for land and desertification. According to the United Nations Convention to Combat Desertification (UNCCD) in 1994, the definition of desertification is land degradation. Many researchers have reported that unsustainable cultivation, over-grazing, firewood collection and urbanization are the major causes of desertification (e.g. Ayantunde et al. 2000, Dregne 1986, Gonzalez 2001, Mortimore and Turner 2005, Tschakert 2007). In recent years, both farmers and herders have experienced hunger and poverty caused by desertification in the Sahel region. This has fuelled armed conflicts. There is a downward spiral of desertification, hunger and poverty, armed conflict and terrorism occurring throughout the region.

In the Republic of Niger, President Mamadou Tandja made a commitment in the beginning of the century to tackle desertification. Government policy has promoted afforestation, water basin management and erosion prevention. According to government reports, during the three years from 2000 to 2002, an area of 381 km^2 was afforested, sand dune fixation has occurred over 40 km^2, and erosion prevention measures have been extended over 384 km^2.

The Great Green Wall for the Sahara and the Sahel Initiative is now being implemented across the Sahel region, with more than eight billion dollars promised in support from the European Union, World Bank, Food and Agriculture Organization (FAO), UNCCD, Global Environmental Facility and other organizations. There are some 20 countries involved in the initiative, including Benin, Burkina Faso, Chad, Djibouti, Eritrea, Ethiopia, Ghana, Mali, Mauritania, Niger, Nigeria, Senegal, Somalia, Sudan and Gambia. The scale of the project is enormous, with the aim of creating a 15 km wide green belt extending 8,000 km across the continent. In Senegal, 11.4 million trees have already been planted on 25,000 hectares of land as part of the initiative.

There is a need to evaluate the effectiveness of this enormous project in providing funds to the Sahel countries through international development organizations. The aim of the initiative is to strengthen the coping abilities of the government and individuals, but there have been difficulties in implementing the initiative. In the Sahel, the population density on arable land is high, with little space available for afforestation and re-greening. Trees can be grown on land that is suitable for agriculture, but it is extremely difficult to plant trees on degraded land without soil improvement efforts. Often forests in the Sahel zone are not natural but are planted or fostered by people for centuries, often in the context of high population densities and specific property institutions, rules and regulations regarding ownership, maintenance and use (see Fairhead and Leach 1995, Haller 2003). However, this requires hard work and large inputs based on many gains (shelter, food, timber and non-timber products, ritual sites, etc.). This shows there has been a willingness to create forests, but it is unclear if under the new conditions and under changed property rights over land on fertile ground people are still willing and able to repeat those efforts.

Leisinger et al. (1995) emphasize the importance of communication between local residents and scientists to provide food security and promote rural development in the Sahel. However, the one-sided transfer of knowledge and skills from scientists to local residents is often not desirable and not adequate, as scientists frequently overlook local knowledge and capabilities in contemporary livelihoods and agricultural practices in rural villages. The local people, especially herders, are framed by the large-scale afforestation project as "troublemakers", and pastoral commons are alienated in the form of large-scale land acquisition or land grabbing, including water grabbing and pasture grabbing (Haller et al. 2016). In rural communities of southern Niger, land shortages have made it difficult for farmers to maintain fallow land. Livestock dung is an important resource to improve soil fertility, but farmers can rarely get enough access to the resource. It is imperative before making expensive agricultural interventions that the concerns of local residents about trying to tackle desertification and overcome hunger and poverty are taken into account.

3.3. Approach and research area

My main exploratory research methodology was participatory observation, which is a widely used anthropological research method (see Crang and Cook 2007), combined with field trials. I started my research in Niger in 2000; it continues today.

When I visited Niger for the first time in 2000, I was surprised by the heat and felt it would be difficult to conduct long-term fieldwork. For me, it was impossible to start research in this arid region without drinking large amounts of water. In the surroundings of the Sahara Desert, we also faced security risks related to Tuareg movements. I intended to conduct fieldwork among the Hausa, the largest ethnic group in West Africa, and decided to establish my research area in Dogondoutchi, 270 km east of the capital city, Niamey (Figure 1). I engaged in agriculture with Hausa farmers and grazed livestock with Fulbe herders.

Figure 1: Research area: D village

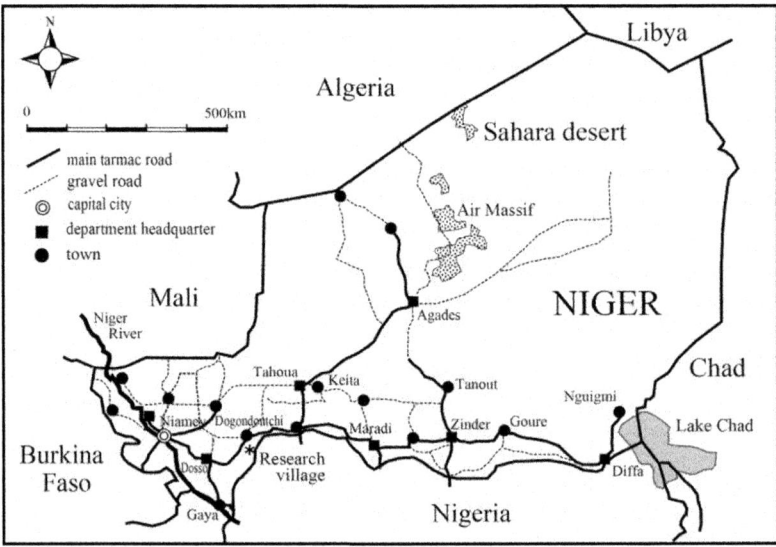

In 2001, the population in Dogondoutchi Town was 29,200. It takes a full day to travel from Niamey to Dogondoutchi by taxi brousse. My research site was a nearby village, which I call D village. In 2000, there were 280 residents among 41 households in this village, which increased to 65 households with 504 residents over the following ten years. During this period, the annual rate of population increase was extremely high at 6.0 per cent. This rate of increase would lead to the population doubling every 12 years. In 2010, 62 of the households were Hausa farmers, two were Fulbe and one was Tuareg. Both the Fulbe and Tuareg are herders of grazing livestock.

The Hausa are the largest ethnic group in West Africa. Hausa people dwell in an area extending from Filingue, Tahoua and Zinder in central-north Niger to as far south as Kano and Zaria in northern Nigeria.[1] They refer to themselves as *Bahaushe* (singular) or *Hausawa* (plural). Major Hausa communities have formed in northern Ghana and northern Cameroon, and Hausa storeowners and artisans can be found in most of the cities of the Sahel. Most Hausa are Muslims, and it is possible to trace the history of this religion in Hausaland back to the fourteenth century. Islam spread as the Hausa migrated from one part of West Africa to another. In West Africa, Islam is now sometimes considered synonymous with the Hausa people (Adamu 1978). The Hausa kingdom flourished as a centre for intermediary trade in produce from the coastal regions to its south, and for rock salt from the Sahara Desert and goods imported from the Mediterranean (Baier 1980). The Hausa actively pursue commerce in urban areas of West Africa.

D village is one of the nine oldest villages in the Department of Dogondoutchi. Elderly villagers recall that, 60 to 70 years ago, spare land for cultivation remained in the plain land around the village, much of it untouched. Aerial photographs taken by the French colonial government in 1950 confirm this, showing the village surrounded by uncultivated and fallow land (Oyama 2017). Fulbe and Tuareg herders grazed their livestock in the grassland of these lands. As the population grew, however, increasing areas of land were cultivated, and by 2006, all arable land had been cleared for farming and the ownership of each plot in the plain had been established. Communal grazing areas have been limited to the iron-crusted inselbergs and their barren surroundings. The paths to the water pond were secured by the government, but the expanding farmlands made it difficult for herders to connect with the water.

---

1   Heiss (2015) provides an excellent ethnography on the everyday life in rural Hausa community.

In recent years, land suitable for cultivation has been in short supply, and it has been difficult for villagers to find newly cleared land. Village women have even attempted to cultivate groundnuts and Bambara nuts in the small sand dunes they have found on the inselbergs. Allowing the land to lie fallow is no longer an option, and pearl millet and cowpeas are grown in succession each year. Much of the arable land is already cultivated and land shortages are serious.

Desertification, food shortage, famine and armed conflicts among farmers and herders are problems that the villagers face in their everyday lives. This article details the environmental conditions in the research area and illustrates the local recognition and management of desertification from the viewpoint of the inner community of the village. At the same time, the paper discusses the methods used for land restoration, improvement of food production, and conflict prevention based on local ecological knowledge and institutional arrangements. The article reveals that counter-intuitive explorations and reverse thinking can provide room for manoeuvre in highly complex situations, but the dilemmas of trade-offs between conflict prevention, food production and environmental degradation remain.

3.4. Agriculture in long-term dry season and short rainy season

3.4.1. Temperature, rainfall and wind

The rainy season in D village is from early June to early October, and is called *damana* in Hausa. The temperature and rainfall in 2010 are shown in Figure 2. The total precipitation in that year was 524.5 mm. The national meteorological station in Dogondoutchi started taking measurements in 1923, and reported an average annual precipitation of 465 mm during the 30 years from 1981 to 2010 from the records of Météorologie Nationale du Niger. The precipitation in 2010 was therefore above the average.

The dry season spans eight months, from October to May. The monthly average maximum temperature usually exceeds 30°C all year round. In 2010, there were 352 days exceeding 30°C out of which 241 days exceeding 35°C. From March to June, the minimum temperature was higher than 25°C during the hot dry season called *rani*. The humid season between May and mid-June is called *bazara*. In the *rani* season, the maximum temperature exceeded 40°C and the minimum temperature around sunrise was still higher than 30°C. For

Figure 2: Temperature and rainfall in D village in 2010 with the seasonal taxonomy according to the Hausa people

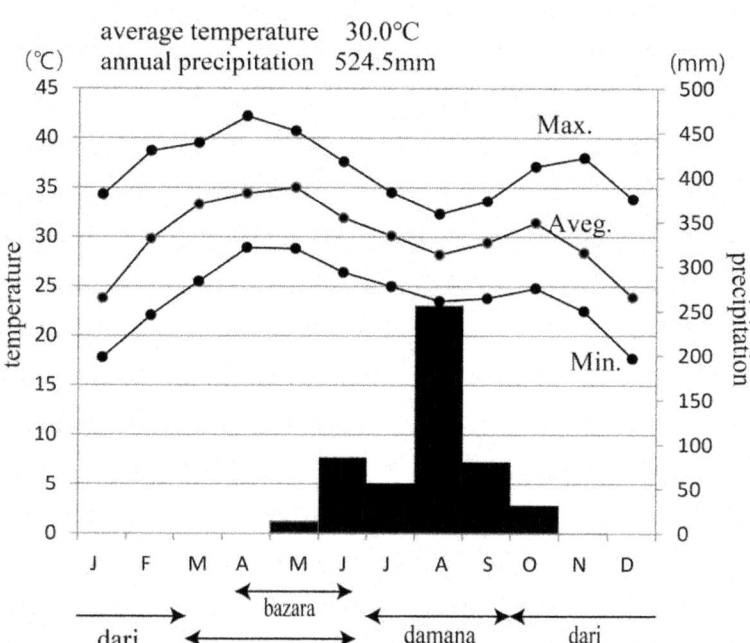

local people, too, these temperatures are difficult and they report that they do not sleep well at this time.

Immediately before it rains, a wind stronger than 10 m/s blows and creates windblown sandstorms. In 2011, there were 95 days where wind blasts of stronger than 10 m/s were recorded. This consisted of 35 days during the rainy season from June to September and 60 days during the dry season. The strong winds usually stop after 10 to 30 minutes. In the rainy season, it sometimes starts raining after the strong wind blows, but this does not always happen. The rain clouds and sandstorms originate from the east and south-east. A hot and dry wind called the Harmattan also blows from the east during the dry season. This wind simultaneously induces soil erosion for land degradation and soil accumulation for land restoration in the Sahel.

3.4.2. Agriculture

There are three staple food crops cultivated in D village: millet, sorghum, and maize. Intercropping fields of millet (*Pennisetum glaucum*) and cowpea (*Vigna unguiculata*) are spread around the village. The millet plant stands straight and the cowpea crawls along the ground. Intercropping of the staple food crop and local legume is almost universal throughout tropical Africa and brings benefits to crop production and soil fertility (Richards 1985). The distance between millet plants is 1.2-1.5 m, with cowpea planted in between the millet plants (Figure 3). The villagers also utilize cowpea in cooking their staple food (*tuwo* in Hausa). Because of the shortage of rainfall, they cultivate mainly millet, and only rarely sorghum and maize. Gourd (*Cucumis melo*), watermelon (*Citrullus lanatus*) and okra (*Abelmoschus esculentus*) are also planted on the farmland.

Figure 3: A millet field in southern Niger

The growing period for millet is three to five months. The exact period is dependent on the rainfall conditions and millet variety. There are four varieties of millet: *zongokolo, dogohatsi, bazaumi* and *maiwa*, whose yield from highest to lowest follows that order. According to the villagers, *zongokolo* requires the highest soil fertility. Drought tolerance from strongest to weakest follows the order of *maiwa, dogohatsi, zongokolo* and *bazaumi*. The variety with the strongest drought tolerance, *maiwa*, is a late maturing variety and has not formed a panicle by the middle of September when the villagers begin to harvest the other varieties. They start harvesting *maiwa* at the beginning of November. The grain of *maiwa* is toxic and the poison has to be removed by soaking the seeds in water. After soaking in water, the grain tastes the same as the other varieties. The use of four varieties creates crop diversity and has enabled the villagers to respond to climate change, but the recent introduction of improved new varieties and chemical fertilizer has led to the extinction of local varieties. This will make local farmers more vulnerable and less resilient to climate changes and drought.

Millet is usually harvested from the middle of September to the end of November. The panicle is cut using a knife. The villagers bundle the harvested panicles together with ropes and store them in a granary or in clay-made storage vessels. Cowpea pods are harvested by all family members, including women and children. Cowpea leaves are collected for use as livestock fodder and stored on the roofs of the houses. Recently a shortage of livestock fodder became a serious problem, and millet stems and leaves have therefore been collected for livestock fodder since 2010.

Men cultivate millet and cowpea in the fields and women plant groundnuts (*Arachis hypogaea*) and Bambara nuts (*Vigna subterranea*) in sand dunes on the inselberg. Women care for small patches of their crops every day during the rainy season. These nuts are important as a household food and also provide cash income for the women. The women also cut trees and collect firewood from the inselberg. The sand dunes are covered with grass, which provides important fodder for the herders' livestock, with most sand dunes cultivated by Hausa women. Deforestation is rapid on the inselberg, and the scattered small patches of nut fields have led to livestock-induced crop damage, resulting in conflicts over land between farmers and herders.

3.4.3. Soil properties and land degradation

The Hausa farmers recognize changes in the soil conditions resulting from continuous millet cultivation (Oyama 2009). For the same general area of land, Figure 4 shows that differences in soil condition and crop production depend on specific locations. The farmers classify the soil conditions into one of three categories: *kasa*, *leso* and *foko*. The Hausa word *kasa* has various meanings, including land in general, country, Hausa land, ground and soil, but here I use its limited meaning of fertile soil. In local conversations, people usually use the phrase *kasa taki* (soil of manure) to refer to fertile soil.

Figure 4: Millet field and stages of degraded soils

There are differences in the soil condition and crop production that depend on specific locations. Left: After a few years of continuous millet cultivation, the fertile soil of *kasa* becomes degraded to *leso*, which appears as white sand in the early stages of land degradation. Right: The solid sedimentary layer, *foko*, is exposed when surface soil is blown away by wind and eroded by rainwater. *Foko* has extremely low plant productivity.

Soil with high organic matter results in high crop productivity. The Hausa call this soil type *kasa*. Compared to the *leso* and *foko* soil types, the *kasa* layer (brownish grey on the Standard Soil Colour Chart), which is 0-3cm and 3-12 cm deep, is weakly acidic and has an abundance of soil nutrients with a rich aggregate structure (see Table 1, Oyama 2012).

This aggregated soil structure is created by termites. Termites use organic matter such as tree branches and leaves as shelter. Their shelters are made of sand held together with saliva and excrement (Lee and Wood 1971). Termites bond the sand grains together with their saliva. Soil aggregates containing air and water promote plant growth. A solid sedimentary clay layer (dull orange), identified as *foko*, accumulates under the *kasa taki* layer. This *kasa* soil type provides favourable growth conditions for millet, with an average stem height of 156 cm recorded on 20 August 2003. The millet crop yielded 1.1 tons/ha in the middle of October 2003.

Table 1: Land and soil classification of Hausa farmers and the soil properties

	pH (H₂O)	Total (g kg⁻¹) N	C	C/N	Exch.Base cmol(+)/kg Na⁺	K	Mg²⁺	Ca²⁺	P Mg kg⁻¹	Soil colour	Sand	Silt (%)	Clay
1. *kasa* (millet yield 1.1 tons/ha)													
0~3cm (*kasa taki*)	6.8	1.20	16.17	13.5	0.06	0.37	2.19	4.36	153	5YR6/1 (brownish grey)	91.0	1.5	7.5
3~12cm (*kasa gara*)	4.8	0.12	1.28	10.7	0.09	0.24	0.09	0.20	8	5YR7/4 (dull orange)	84.2	1.5	14.4
12~30cm (*foko*)	4.4	0.08	0.84	10.5	0.02	0.04	0.04	0.09	6	5YR7/3 (dull orange)	84.6	1.3	14.1
2. *leso* (millet yield 0.1 tons/ha)													
0~9cm (*leso*)	6.1	0.07	0.75	10.7	0.02	0.07	0.09	0.25	7	5YR8/4 (pale orange)	94.6	1.0	4.4
9~30cm (*foko*)	4.6	0.11	1.18	10.7	0.02	0.1	0.061	0.13	5	5YR7/4 (dull orange)	90.5	0.6	8.9
3. *foko* (no millet yield)													
0~5cm (*foko*)	4.6	0.12	1.08	9.0	0.01	0.10	0.12	0.26	13	5YR6/4 (dull orange)	89.5	2.2	8.3
5~30cm (*foko*)	4.4	0.08	0.84	10.5	0.01	0.08	0.05	0.14	7	5YR6/4 (dull orange)	82.0	2.4	15.6

The *kasa* soil type changes into *leso* after a few years of continuous millet cultivation without a manure input. This *leso* soil type shows an early, degraded soil condition that produces low millet yield. The *leso* soil type has a poor aggregate structure and is a pale orange sandy soil that contains little silt and clay. Under the *leso* topsoil, a sedimentary layer of *foko* forms, which appears as dull orange sandy soil. This sandy *leso* soil has high soil porosity and does not disturb the root growth of the crop, but nutrient availability is poor. The average stem height of millet in the fields with *leso* soil was 36 cm on 20 August 2003. The plants failed to form panicles, and the grain yield was only 0.1 tons/ha.

A few years of continuous cultivation without land management leads to wind and water erosion of the topsoil and exposes the solid sedimentary layer. This sedimentary *foko* layer has extremely low plant productivity. Hill (1972) describes this hard and barren ground in his excellent ethnography of the Hausa people in northern Nigeria. The *foko* layer is mainly a quartz sand containing acidic sulphate. The *foko* soil has strong acidity and poor soil nutrition. The clay layer is runny when wet, but hardens after it dries. When the *foko* layer is exposed at the surface, crop growth at the root is significantly hampered due to the soil's single-grain structure and poor chemical constitution. The solid *foko* layer greatly impedes water penetration into the ground. The millet germination rate in this soil is low, and most plants eventually die. All the millet had withered by 20 August 2003, with an average stem height of only 7 cm. The millet grain yield was zero (Oyama 2009).

As Graef and Haigis (2001) note, the soil fertility of farmland is decreasing and further technologies need to be integrated into farming systems to make them sustainable in the Sahel. From the viewpoint of Hausa farmers, land degradation means a transformation of the ground surface from *kasa* to *leso* and then from *leso* to *foko*. According to their explanation of the land degradation process, *kasa* will change to *leso* and then *foko* without land management. In the first step, soil fertility is lost with the change to *leso*. Then the white sand of *leso* is blown away and eroded by rainwater until a solid sedimentary layer of *foko* is finally exposed at the surface.

This land degradation process is caused by water and wind erosion. Surface soil is eroded away and a solid sedimentary layer is exposed to the surface. This process is triggered by anthropogenic factors such as continuous farming, grazing and tree-cutting. Loss of the surface soil creates a barren land surface of sedimentary rocks, which are called *foko* by Hausa farmers. This barren land is the final stage of the land degradation process. The villagers cannot produce millet and cowpea on this land, and natural plants cannot become established.

Initiatives like the Great Green Wall for the Sahara and the Sahel Initiative emerge from such environmental analyses to counteract desertification and to restore land for agricultural production and to sustain livelihoods.

3.5. Local countermeasures against land degradation

The Hausa farmers do not simply accept land degradation. In Hausa society, the importance of *harkuki* (movement) is culturally emphasized. Farmers respond rapidly to the degradation of their farmland. They proudly told me: "We can improve the soil condition of *leso* and *foko* and convert it back to *kasa*." To recover the crop yield, they have adopted the two countermeasures: (1) providing livestock dung to farmland based on an encampment contract with herders in areas of *leso* soil, and (2) providing waste inputs to the degraded *leso* and *foko* soils.

The encampment contract in the Sahel region allows maintenance and improvement of the productivity of millet plots. These institutional arrangements between farmers and livestock herders, called *hulda da makyaya* in Hausa, have been widely practised for generations across semi-arid West Africa (e.g. Baier 1980, Schlecht and Buerkert 2004).

With the arrival of the rainy season, the nomadic Fulbe and Tuareg people move north, crossing the marginal limit of the cultivation zone and grazing their livestock around the edges of the Sahara Desert. The Fulbe people mainly raise cattle, together with goats and sheep, while the Tuareg herders may have camels as well as goats and sheep. Women and children move from one camp to another on donkeys, which are also used to transport household possessions.

In November and December, as the rainy season comes to an end, the herders begin to move their livestock south, passing though farming villages. By this time, the villagers have finished harvesting their millet and cowpea, and the land surrounding the village is covered in post-harvest stubble. The herders try to select a route that will take them through the most productive farmland, and approach villagers - particularly wealthy ones - to discuss encampment contracts.

Once the parties have reached an agreement on terms such as the camp duration and fees, the herder sets up his camp on the contracted land. During the daytime, he lets his animals feed on the post-harvest stubble, leaving them free also to graze on plots other than the one subject to the contract. In the evening, he brings them back to the campsite, where they spend the night. The livestock are not corralled, but the herder keeps livestock on the contracted land. The animals nourish the soil in the vicinity of the camp with significant quantities of manure (Figure 5). The addition of these nutrients is crucial to the outcome of the following season's millet crop (Schlecht and Buerkert 2004, Shinjo et al. 2008, Suzuki et al. 2014). At the end of the contract period, the herder receives payment from the farmer in cash and millet, and leaves the village.

Figure 5: Encampment contracts between farmers and herders

Farmers make encampment contracts with the Fulbe and Tuareg herders. During the night, the herders stay with their livestock on the farmer's farmland. The farmer exclusively receives soil nutrients on the private farmland from the livestock dung. Farmers usually request that herders camp on the sandy *leso* soil, which is nutrient deficient.

The fertilizing effect of the manure from these encampments is encapsulated in the Hausa saying: "Camel five years, goat and sheep three years, cow one year". Cow manure decomposes rapidly, providing nutrients for one year only; goat and sheep manure can last for three years, while camel manure provides five years of fertilization. The Hausa farmers' understanding of the properties of livestock manure is consistent with the scientific findings of Brouwer and Powell (1998). Villagers wishing to enter into contracts with the Tuareg, whose herds include many camels and therefore have the potential to provide highly effective fertilization, are required to pay a high price in cash and millet. Soil nutrients can be added to the degraded farmland through this encampment contract.

3.5.1. "Waste is manure for our farmland"

Hausa farmers have another countermeasure to restore the degraded farmland. As part of their everyday life, villagers collect waste to fertilize their own farmland (Oyama 2012, Oyama and Mammane 2010). Waste is regarded as manure, called *taki* in Hausa. The famers monitor the soil conditions on their farmland and apply the waste to areas of degraded land (i.e. *leso* and *foko*) as a land management activity (Figure 6). They emphasize the importance of waste inputs and the biological activities of termites for recovering crop productivity. I have conveniently translated the Hausa word *taki* into manure in English, but *taki* means not only organic waste (e.g. the inedible parts of crops, such as plant residues from food processing, and livestock fodder and dung) but also inorganic waste (e.g. worn out clothes, shoes, vinyl sandals, plastic bags, cartons, straw baskets and mats, iron pots and dishes, and used batteries). The villagers claim that non-compostable waste is important for fertilizing the farmland. The farmers transport domestic waste from their homes to their farmland throughout the year. The amount of waste produced varies according to the number of household members and livestock, and the animal species, but the average amount is 10 to 40 kg per day. From a household owning an ox cart, 200 to 400 kg waste is transported to farmland every seven to ten days. Men and women wrap 10 to 15 kg of waste in cloths and carry it to the farmland on their heads every day.

Some farmers living near the town collect urban waste for application to their farmland, with the aim of improving the soil condition. They transport urban waste to the degraded *foko* land by ox cart, tractor or dumping vehicle. Urban waste contains large amounts of soil nutrients such as nitrogen, potassium, phosphate, calcium and magnesium. Sand accounts for at least three quarters of the mass of urban waste and is important for land restoration.[2] This sand is carried by storms immediately before the rainfall and Harmattan during the dry season.

---

[2] According to my weight measurement in February 2018, the urban waste of one tractor was 2826 kg in total. The composition was 2609 kg (92.3%) sand, 189 kg (6.7%) plastic and vinyl bags, 17.4 kg (0.6%) stone, 4.96 kg (0.2%) plastic, 2.48 kg (0.09%) metal, 2.26 kg (0.08%) cloth, 0.68 kg (0.02%) glass fragments and 0.11 kg (0.004%) paper. Even in the urban area, a huge amount of sand is provided by strong wind and sandstorms. The urban dwellers collect the sand and carry it to the dumping site. This sand disturbs the incineration treatment of urban waste in the Sahel.

Figure 6: Waste for fertilizing the farmland

Farmers monitor the soil conditions and apply urban waste as manure to fertilize the degraded farmland. (A) Farmers put millet stems and plant residues onto the degraded land. (B) Sand accumulates around the livestock dung, plant residue and pruned branches. (C) In 2010, farmers started using the urban waste to catch the blown sand. (D) Farmers scattered the waste by fork and promoted termite activities and organic decomposition. The urban waste includes plastic bags, worn out clothes, sandals, baskets, and metal dishes and pots. Farmers claimed that all the materials have functions and good effects for soil improvement and land rehabilitation.

The organic matter provides nutrition for crops and food for termites, so improving the physical condition of the soil. The inorganic matter in worn out clothes, sandals, mats and pots is mixed in with the blown sand in the urban waste and accumulates in the sedimentary layer. The urban waste could be considered a soil dressing applied to the degraded land. Applying urban waste can help to recover the plant productivity of degraded land (Oyama 2012, 2015a, Oyama and Mammane, 2010).

3.5.2. First trial of urban waste-induced land restoration

In a meeting in June 2005, I consulted with the village headman and influential seniors about my plan for a land restoration experiment using urban waste and requested permission to use the degraded land to the east of D village, surrounding an inselberg. All of them approved my request willingly and promptly. At that time, an aged man talked about past activity in this area. He said:

> "The degraded land you want to use was pasture land more than 60 years ago. The ground was covered with grass and a Fulbe herder named Boi lived with his livestock on the grassland. After the trees were cut down for firewood and Boi grazed livestock continuously for many years, the wind blew away the sand and the soil was eroded by rainwater. The erosion exposed the barren rock and plants could no longer grow, leading to the current situation. The ground became unsuitable for grazing livestock and Boi left the village with his son, Madaure."

The ground was "no man's land", with free access to the public. The herders could use the grassland freely and induced the land degradation.

The headman and influential seniors gave me permission to use an area of 2.7 ha. The land was positioned on a gentle slope, and included the pediment around the inselberg. Rainwater flow had washed away the surface soil and the pediment was vulnerable to further water erosion. The sedimentary layer was exposed on the surface and no plants were growing on the barren land. In 2007, it was estimated that between 80 cm and 1 m of soil erosion had occurred by measuring the length of exposed tree roots (Figure 7). The solid, barren ground was classified as degraded land, *foko*, by the farmers. There was no grass and a limited number of trees in the area. We could not find any active termite mounds. Only one old mound was observed in the area; it had been weathered by wind and rainfall (Figure 8).

I rented two bulls and ox carts to carry urban waste from Dogondoutchi to the degraded land. The rental fee was 1,500 CFA francs (3 USD) per journey. It took one and a half hours each way. When we arrived at the town, we collected the waste from a residential area using rakes and loaded the carts with waste using shovels. Under the strong sunshine, I developed blisters on my right hand due to the unfamiliar and strenuous work. At the beginning of June, in the early rainy season, the urban waste was moist and had an unpleasant odour.

Figure 7: Tree roots exposed by wind and water erosion, indicating soil erosion to a depth of at least 80 cm

At first, the residents were perplexed to see us working in the town and transporting waste to the village. I explained the purpose of the experiment to the urban residents and they immediately understood and thanked us for cleaning the residential area and surrounding streets. After loading each ox cart with waste, the carts became heavy and it took more than two hours to reach the site. The two bulls were salivating and struggled to pull the carts.

When we arrived at the experimental site, we dropped the waste from the ox carts and levelled the waste into a round shape. The mass of waste was 400 kg per ox cart. The bull owner complained about the low price for the rental fee and demanded an increase from me. I agreed to increase the fee to 2,500 CFA francs per journey. July is considered the hungry season, when people suffer food shortages. It was also a hungry season for the livestock: we could see the bulls' ribs, and they did not have enough power to pull the ox carts when they were fully loaded with waste. Although I increased the rental price

Figure 8: Weathered termite mound

In this arid land, the most common organisms in the underground environment are termites, but many termite mounds have been abandoned and have subsequently become weathered because the termites cannot get access to wooden material as food following desertification.

to 2,500 CFA francs, we had to reduce the number of journeys made by the bulls to 12.

Some villagers asked to borrow money to purchase household food at that time, but I refused to lend money to avoid later money troubles. I could not reject their requests without being impolite and so, instead, I requested they bring domestic waste to the site for which I paid them a fee. I was able to collect waste 12 times by ox cart, ten times on the back of a donkey, and once carried on a man's head. A total of 5,200 kg of waste was applied to eight locations on the degraded land (Figure 9).

Figure 9: Land restoration experiment using urban waste

3.5.3. Emerging pastureland

I visited the research site in August 2008, which was in the rainy season, two years after we carried the urban waste to the degraded land. I observed a green carpet that had arisen from the urban waste. The green colour was in stark contrast to the surrounding degraded land, which had a uniform brown colour. The son of the village headman was experienced at identifying plant species and counted 75 species at the site (Figure 10). According to my Hausa informants, plastic bags, sandals and metal pots were able to catch the wind-blown sand and so prevented soil erosion. My Fulbe informant explained that most of these plants growing on the urban waste had high levels of nutrition and were favoured fodder for livestock. We collectively believed that urban waste could be used to restore the degraded land. My informants said: "Both urban and domestic waste can be used to create pastureland." Interestingly, the emerging pasture was immediately used by more people in the village.

Figure 10: Grassland created from urban waste

I encountered a Tuareg girl in the pastureland created from the urban waste. She grazed ten goats and one sheep for her parents. I knew her parents and they all lived in D village. The time of this encounter was 1:30 pm and the site was in full sun. Despite the heat, she continued to chase and graze the livestock on the pastureland. The girl said to me:

> "If I return home now, there is nobody at home. My mother and father are working on the farmland. I feel good grazing livestock here. When I graze them from the early morning, we can get milk from the sheep and goats in the evening and the milk nourishes us. But Hausa farmers will scold me terribly if I enter their farmland. I am happy to graze here, away from the farmland."

After my conversation with her, I was able to understand more clearly the severe situation faced by Fulbe and Tuareg herders. They needed to secure pastureland to graze their livestock freely without intervention from the farmers (Figure 11). However, free access to pastureland promoted further land degradation. Although I had made a huge effort to transport the urban waste and restore the land, the livestock grazed the pastureland freely and removed the soil nutrients. A few years later, the desertification process would start again, with conversion back to the degraded land referred to as *foko*.

Figure 11: A 12-year-old Tuareg girl grazing goats and sheep on the grassland created from urban waste

3.6. Eight effects of urban waste use for land restoration

It became clear that land restoration from degraded land to pastureland was possible technically. However, I could neither understand the process and mechanism of land restoration nor the necessary amount of waste for land restoration. In November 2008, I consulted the headman and influential seniors again about the possibility of building a large-scale experimental plot. Then I started another trial, as I wanted to assess the effects of using urban waste for land restoration.

With the cooperation of the villagers, we transported urban waste from a residential area of Dogondoutchi for use in a greening experiment on the sedimentary layer. We enclosed the land with fences for 50 m in the east-west direction and 45 m in the north-south direction (Figure 12). The experimental site was located on the pediment, with a gentle slope, and was highly vulnerable to soil erosion. There was no plant growth around the plot and we did not observe any active termite mounds nearby. This area of pediment was the location that the seniors had informed me earlier was degraded pastureland. The termite mounds had been diminished by rain and wind, and tree roots were exposed to a depth of 30 to 70 cm.

In this 50 × 45 m plot, I made five sub-plots (plots 1 to 5) that extended 4 m in the north-south direction and 30 m in the east-west direction. At 20 m from the east end of the plots, I set two time-domain reflectometry (TDR) sensors for soil moisture at depths of 5 and 20 cm in each plot. These ten sensors measured soil moisture every hour and the data were recorded by a data logger. The temperature, wind direction and speed, and rainfall were also measured and recorded every hour. A technician from the Météorologie Nationale du Niger generously supported my work.

We hired an open Toyota Hilux to carry urban waste from Dogondoutchi town. We measured the weight of the urban waste using a scale (LDS-30H, Shimadzu, Tokyo, Japan; minimum scaling unit was 0.01 kg). Plot 1 was a control and did not contain any urban waste. We placed 5 kg/m^2 (total 600 kg) of urban waste in plot 2; 10 kg/m^2 (1,200 kg) in plot 3; 20 kg/m^2 (2,400 kg) in plot 4; and 45 kg/m^2 (5,400 kg) in plot 5. We placed waste into a flat iron pan and measured the amounts with the help of 15 villagers. One villager levelled the waste on the ground with worn out sandals and said to me, "This trial is very good, and many plants will grow from the urban waste. I will buy this land for sowing millet." He said this to me repeatedly.

The in-situ experiment revealed that urban waste inputs to degraded land improve plant growth through a combination of the eight factors described below (Figure 13). The arenosol soil type[3] is prone to damage from water and

---

3   The arenosol soil type - one of the typical poor soils in Africa - is also found in Namibia. According to Prudat et al. (2018), local farmers recognize this soil type as *Omufitu* and its suitability for pearl millet is poor. Prudat et al.'s paper integrated the farmers' assessment and technical knowledge in order to develop the practical soil quality assessment.

Figure 12: Experimental plots two years after being established (August 2010)

(a) Plot 1: control plot, no waste input. (b) Plot 2: urban waste input 5 kg/m^2. (c) Plot 3: urban waste input 10 kg/m^2. (d) Plot 4: urban waste input 20 kg/m^2. and (e) Plot 5: urban waste input 45 kg/m^2

wind erosion (Bleich and Hammer 1996), but low mounds with a range of elevations superimposed on a flat topography can trap sand and organic matter that is blown by the strong east winds (first effect of Figure 13). This is the same technique that Michels et al. (1995) used to alleviate the effects of wind erosion using millet residue, but crop residues do not usually remain in the field in this region because of livestock grazing and termite decomposition. Local people collect crop residue and take it to their homesteads, where it is eaten by livestock. The Hausa people welcomed the addition of plastic sandals, bags, metal pots and plates in the waste scattered onto their fields: because these items do not easily decompose and are not affected by the termites, they cover the soil and trap windblown sand for longer than organic waste. These wastes prevent erosive wind and water (second effect).

Figure 13 Eight effects of urban waste inputs used for land restoration

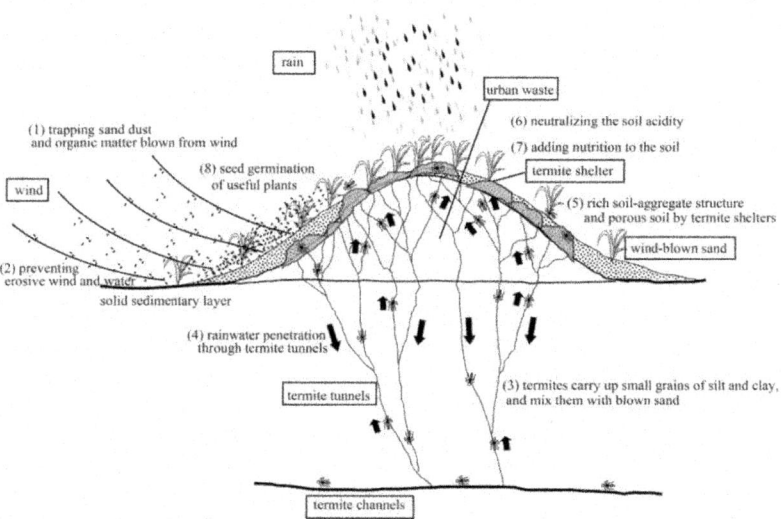

I also considered the various effects of elevated termite activity due to the waste input. Most of the waste consisted of millet stalks and leaves, leftover livestock feed and animal excreta. Waste inputs encourage termites to gather. Termite guts and nests contain symbiotic microorganisms such as bacteria, protozoa and fungi that decompose cellulose and lignin, fix nitrogen and produce methane (Benemann 1973, Lee and Wood 1971). These biological activities alter the chemical properties of the soil and, as a result, termite mounds can contribute to high levels of soil fertility (Adepegba and Adegoke 1974, Bagine

1984, Benemann 1973, Pomeroy 1976). As a result of this termite activity, termite shelters develop over the organic matter, these containing concentrated amounts of organic matter. Termites also dig up small grains of clay and silt particles in the soil and mix them with windblown sand (third effect). Termite tunnels also penetrate the sedimentary layer, allowing rainwater to infiltrate easily through the tunnels (fourth effect), and an aggregated soil structure is created as the termites stick grains of sand together with their saliva as they build their mounds. Our observations showed that the aggregated soil structure was porous, allowed plant roots to grow and easily penetrate the soil, and contained oxygen and moisture, both of which are necessary for plant growth (fifth effect).

These factors ameliorate the poor natural nutrient content and strong acidity, as indicated by the low pH of the parched and degraded land. Organic matter, including livestock excreta, contains large amounts of nitrogen, phosphate and potassium, and significantly improves the chemical properties of the soil. Urban waste and excreta are neutral to alkaline, and neutralize the soil acidity (pH 4.5) of the degraded land (sixth effect), as well as adding nutrients to the soil (seventh effect).

Finally, urban waste contains seeds of many edible plant species, including millet, *Hibiscus sabdariffa*, *Balanites aegyptiaca* and many other plants that are suitable as feed for livestock (eighth effect). In Sahel, the seeds of the predominant crop and fodder grass are very small and easily mixed in the urban waste. These germinate naturally with the arrival of the rainy season; in the experimental plots we set up, the seeds germinated and grew thanks to the presence of the moisture and nutrients derived from the waste.

The eight effects described above can be combined to improve soil fertility and plant growth productivity. According to the estimations of the villagers, plant growth in plot 3 was not sufficient, but in plots 4 and 5 growth was sufficient to provide fodder for livestock. In conclusion, the minimum amount of urban waste that should be applied to degraded land to ensure plant growth productivity is 20 kg/m^2, a depth of 3 cm. However, after three years of waste input without further land management, land degradation began again.

3.6.1. Safety issues with urban waste

I have frequently received questions, comments and criticism regarding the toxicity of urban waste following its application in the environment. For example, I have been asked if there are instances where livestock have died after eating plastic bags. The animals, including cattle, sheep, goats, camels and donkeys in the Sahel sometimes did eat plastic bags together with fresh food, but they would never eat plastic bags with spoiled food. The animals have a good sense of smell. We usually allowed the livestock to enter the fenced pastureland after a period of more than three months following the deposition of the urban waste. The livestock never ate plastic bags at that time. Anyway, I was strongly conscious of avoiding the risk of contamination.

I examined the safety of urban waste for land restoration practices and analysed the heavy metal concentration of 100 waste samples in the capital Niamey using energy dispersive x-ray spectrometry (EDX-700HS, Shimadzu). Only five samples had harmful levels of heavy metals (bromine, chrome and lead). These samples were collected from long-term waste dumped at the roadside, in street ditches and in an industrial area (Figure 14). The urban waste disposed of directly from homesteads met the safety requirements of European Union environmental standards. We assume that all urban waste will contaminate the environment, but actual urban waste from Niamey city does not contain high levels of pollutants.

Consequently, we can maintain the safety of urban waste for the land restoration project by avoiding the use of long-term waste dumped at roadsides, in street ditches, in the marginalized greenbelt and in industrial areas. I was able to recognize the safety of urban waste which was dumped immediately from the house yards. According to chemical analysis, urban waste contains beneficial materials such as nitrogen, potassium, phosphate, copper, iron, magnesium, calcium, manganese, sodium and zinc, which are essential for humans and livestock (Oyama 2015b).

3.6.2. Collecting waste from the city administration to resolve the financial deficit problem

After a coup d'état in February 2010, a military regime was established in Niger. Soldiers took over the city administration in Dogondoutchi in an operation that lasted until the election of January 2011. I returned to Niamey in October 2011. The next day, I went to Nouveau Marché with my research partner. We bought a 200 m long fence, wire, an entrance door, paint and a brush for construction of the experimental site. We went to Dogondoutchi City Hall to get a permit to collect urban waste. We met the mayor and explained our three objectives: (1) to collect urban waste and therefore clean the residential area and streets; (2) to restore the degraded land back to pastureland in the

Figure 14: Contaminated urban waste at five sites

Pb-1 to Pb-3: Lead was detected at three sites of street ditch and ash; Cr-1: Chrome was detected in long-term dumped waste; Br-1: Bromine was detected from the road-side.

fenced area using the urban waste; and (3) to prevent conflicts between farmers and herders by allowing the herdsmen to graze their livestock on the pastureland during the harvest season. The mayor understood the aims of our project, but explained that there was a budget shortage due to the military regime. There were 28 civil servants working in Dogondoutchi at the time, and they did not receive a salary under the military regime. The waste disposal trucks were not maintained properly and all of them were out of service. During this period, the city was not collecting waste from the urban area and after the election the conditions did not improve due to a financial deficit. The city was unable to cope with the waste accumulation and dirtiness in the urban area. The mayor therefore thanked us for cleaning the city and using the urban waste for land restoration. We made an agreement with Dogondoutchi City Hall, which was detailed in formal documents.

The mayor introduced me to a tractor owner: he was among the wealthiest merchants in Dogondoutchi, owning a hotel, restaurant, several shops, 200 ha of farmland, several hundred cattle, sheep and goats, and a garden of mango trees and vegetables in a wetland area. He also owned dozens of houses that were rented out, and had seven vehicles, including one Toyota land cruiser, one truck and two tractors. He was engaged in the construction of government buildings, including the city hall, an official residence and water wells.

He considered the introduction from the mayor and provided me with two tractors without rental fees. I paid for the fuel and labour costs for waste transportation. One journey from the town to the experimental site cost 8,000 CFA francs (16 USD). The size of the carriage was 310 cm (length) × 180 cm (width) × 60 cm (height). Six young men used shovels to haul the waste into the carriage. I selected waste that had been recently disposed of in the residential area in order to avoid issues of contamination.

I recorded the initial weight of the transported waste. Before letting the 15 villagers begin to work, I requested that they wear masks. The temperature was higher than 40°C and the masks made it uncomfortable to breathe, but the workers were required to wear the masks all the time because strong wind sometimes blew the waste and dust. The first carriage transported 1.80 tons of urban waste, containing millet straw and stems, and pruned branches. The urban waste in the second and third carriages contained large amounts of sand and weighed 3.31 and 3.27 tons, respectively. The tractor driver considered that urban waste containing large amounts of sand would be suitable for land restoration. We calculated the average weight of urban waste per carriage to be 2.8 tons. Because an ox cart can carry 400 kg, this amount was equivalent to seven ox carts.

According to my previous experiment, at least 20 kg/m^2 of urban waste (i.e. the amount used in plot 4) was necessary for land restoration. We estimated that 50 tons of urban waste would therefore be necessary in a 50 × 50 m plot. This amount was equivalent to 18 tractor loads. To use the same amount as that applied to plot 5 (45 kg/m^2), the amount of urban waste required would be 112.5 tons, which was equivalent to 40 tractor loads (Figure 15).

Figure 15: Building an experimental plot for restoring the land by urban waste

I hired twenty village men in order to support their livelihoods. We dug ditches to prevent rainwater from flowing away from the urban waste. To prevent the wind from blowing away plastic bags, we also carefully placed sand and sedimentary rocks on any loose plastic material.

The urban waste contained a range of items including worn out sandals, clothes for women, trousers for men, pots, plastic shopping bags, vinyl bags used to contain milk and mineral water, and empty boxes of tobacco. The bones and dung of livestock were found in the waste, as well as human excrement. These items were suitable for land restoration inside the fence, but were regarded as contamination outside the fence.

The distance from the experimental fence to the village well was more than 1 km and the water depth in the well was more than 40 m. We therefore did not need to worry about the possibility of ground water contamination, but I decided to dig ten long ditches and 20 half-moon shaped ditches to prevent rainwater from flowing away from the urban waste over the land surface.

3.6.3. Inviting livestock into the fenced pastureland

In September 2012, I came back to the village with three graduate students from my university. I went to observe the experimental site as soon as we arrived at the village. The students followed me to the site. In front of us, green plants had spread across the degraded land (Figure 16). We counted 39 plant species, including seven tree species inside the fenced site. In the first year, the dominant species were crops such as millet, sorghum, watermelon, pumpkin and groundnuts, and fodder species of *Poaceae* and *Leguminosae*. Seeds of these species were present in the urban waste. Baobab tree, *Adansonia digitate*, was present as an ingredient in soups and snacks (Figure 17). People eat the raw fruit of *Piliostigma reticulatum* and the leaves and seeds of *Balanites aegyptiaca*. The leaves of *B. aegyptiaca* taste bitter, but are regarded as the "final" famine food by the villagers: they eat this plant for survival in times of famine. Fruits of the date palm are produced near the Sahara Desert and are transported from Agadez to the Sahel towns in Niger. This is a popular food item among urban dwellers and they dispose of the seeds after consuming the fruit (Figure 18).

Figure 16: Plant survey on the first-year plot

Figure 17: A Fulbe woman collecting baobab leaves on the plot

My Fulbe informant considered there to be sufficient fodder for livestock in the fenced plot in terms of the variety and quantity of plants. I became aware that he was reluctant to allow his livestock to enter the fenced site. I requested that he graze his livestock inside the fenced pastureland, and his son brought 40 cows and 12 sheep to the site. After they approached the site, the livestock stopped at the front of the door and hesitated to enter. In a joint effort, we chased the livestock into the site. When the livestock entered the site, they started to panic and run around. After a while, they calmed down and began to eat the grass (Figure 18).

Another Fulbe herder who lived in a neighbouring village asked me, with some hesitation:

> "I cordially request you to build a fenced site in the same way and carry the urban waste to my village. All of the Fulbe suffer from a shortage of pastureland. We are frightened that the farmers will attack us someday in the near future."

He spoke in Hausa with idiosyncratic pronunciation. Nevertheless, I understood his request and made a promise to build the fenced pastureland.

Figure 18: Livestock eating in the grassland created from the urban waste

3.7. Conflict prevention and livestock-induced land restoration

In the Sahel, the number of armed conflicts between farmers and herders has increased in recent years. According to the Global Terrorism Index 2015 (IEP 2015), there were 1,229 deaths attributed to Fulani (Fulbe) militant attacks in 2014, which was a substantial increase from the 63 deaths recorded in 2013. Fulbe communities have come into dispute with farming communities over resources. Fulbe militants have used machine guns and attacks on villagers to assault and intimidate farmers. To achieve regional stability, it is necessary to avoid armed conflict between the farmers and herders and to stabilize the livelihood of the small number of herders (Turner et al. 2011). Most of these armed conflicts have been caused by livestock-induced crop damage (Oyama 2014).

Livestock-induced crop damage has worsened the relationship between farmers and herders in two ways (Oyama 2014). Firstly, careless herders lose sight of their livestock, which then eat the millet in the farmland, leading to non-intentional livestock-induced crop damage (Figure 19). Secondly, there is intentional livestock-induced crop damage, where herdsmen deliberately enter farmland with their livestock and let them consume crops. The farmers are most concerned about intentional crop damage.

Figure 19: Night grazing

Cattle wake up at 10:00 pm every night and start wandering away from the grazing camp. The herdsmen follow the livestock as they start night grazing. To fatten the livestock, night grazing is an important activity. Most of the herders do not have flashlights. They check the position of livestock using moonlight, but they sometimes lose sight of the animals on dark nights.

Many armed conflicts have occurred between farmers and herders following disputes over livestock-induced crop damage. Farmers sometimes use abusive words to the young herders while the animals are grazing. Some of them jump to the conclusion that the livestock must have entered their fields and consumed their crops. Young herders are angered by the insults and fight back against the farmers. Farmers have also physically attacked herders who were intentionally grazing livestock on their farmland.

Due to the rapid population increase, cash economy and market activities, the area of farmland is expanding and the area of pastureland is decreasing rapidly. There has been a persistent misconception of the nature of pastoral economies, combined with increasing land alienation and fragmentation through government policies and privatization of pasture (Haller et al. 2016).[4] At the local level, the herders are finding it difficult to find suitable pastureland. The herders living in the village lost their livestock and their livelihoods due to repeated droughts during the 1970s and 1980s. They now take care of the farmers' livestock and receive a small caring fee and some cows' milk in return. Young farmers are also suffering from a shortage of farmland. The situation is becoming increasingly difficult for both farmers and herders, and the tensions that have arisen have resulted in armed conflict between them.

I was told by one local resident that, "The farmers and herders in the Sahel are friends during the dry season, but become enemies during the rainy season." The armed conflicts are limited to the rainy and harvesting seasons. During this time, the proximity of livestock to farmland is problematic for the local residents. However, the area of farmland and number of livestock have both increased drastically. It is therefore unavoidable that livestock will come into closer proximity with the farmland.

Moritz (2006) suggested that the proximity of livestock to farmland and the subsequent conflicts could be reduced by decentralizing the regional government and authorizing traditional leadership. In the meantime, the number of deaths due to armed conflicts is increasing. However, as Haller (2002) and Haller et al. (2013) show, what is necessary is to find a way whereby the old institutional design can be revitalized, and not decentralization per se, but local involvement in the co-creation of institutions. In addition, a more effective approach is not to rely solely on local authorities but to create a platform for local institutions creating dialogues between farmers and herders. This platform could reduce the proximity of livestock to farmland under these stressful conditions, but large-scale afforestation without social consideration increases the proximity of livestock to farmland and increases the risk of armed conflicts between farmers and herders.

After my field experiment was repeated, I proposed a solution to restore the degraded land and provide pastureland to the small number of herders. According to this plan, we enclosed an area of at least 50 × 50 m with a fence.

---

4   Following the independence of the Republic of Niger, the successive governments and central administrations consisted of the educated, political, urban elite with ambitions of modernization, continuing to reduce the political, social and economic power of the chiefs. After ten years of continued cultivation, the use right was transformed into a property right under the Dirori government regime (1960–1974) (Lund 1998).

The size of the enclosed area was determined by the surrounding landforms, such as the inselberg and the seasonal river or *wadi*. We collected urban waste and transported it to the site. At the end of the rainy season, the farmers start harvesting millet on their farmland; during this period, there is extreme tension between farmers and herders. The herders allowed their livestock to enter the fenced plot at night. During the night, the livestock grazed inside the fenced site and the herders avoided the night grazing of crops. During the day, the herders could allow their animals to graze freely both inside and outside the fenced sites.

However, if the livestock consume all of the grass, the soil nutrition will deteriorate, and desertification will result. I requested that the herders keep livestock inside the fenced site during the night for at least two weeks after they had consumed all of the grass. This enabled the animal dung to provide soil nutrition for the following year's plant growth, as well as avoiding livestock-induced crop damage and livestock theft during the night.

In October 2017, when the system had been in operation for six years, my Fulbe informant said to me: "Urban waste could create suitable pastureland. The cows are satisfied with the fodder and they can relax under the shade of trees. This is the first time we have seen this during the day time." To meet his needs for the creation of pastureland, we had repeatedly removed unnecessary grass, trees and branches during the six years. He also emphasized the importance of avoiding livestock-induced crop damage and armed conflicts with farmers. He was no longer worried about cattle movement at night and could sleep very well until sunrise (Figure 20). He invited his Fulbe friends and Tuareg neighbours to graze their livestock inside the fenced pastureland and welcomed Hausa women collecting the edible plants and medicinal herbs. The movement can be regarded as the establishment of commonality among the multi-ethnic residents. This commonality is informal based on face-to-face relationships, and can be created naturally by the generous-hearted mind of the residents.

Some farmers complained to me that I gave too much attention to the herders at the expense of the farmers. The farmers' concerns were that they were poor and did not have enough farmland to sustain their livelihoods. Effectively, the livestock that the herdsmen take care of every day are mostly the property of farmers. The herdsmen own less than 20 per cent of the livestock they manage. In my Fulbe informant's herd of 40 cows, he owns only two cows, the other 38 belonging to farmers. The fence construction provides benefits to all residents, not only for the Fulbe herders but also for the Hausa

Figure 20: Pastureland created from urban waste in September 2017

We repeatedly removed unnecessary trees and grass to make pastureland according to the requirements of the Fulbe herders. By the sixth year, the plot represented optimal pastureland for the herders. There was no plastic waste visible on the ground.

farmers whose livestock are fattened. The project also had the aim of preventing armed conflicts between farmers and herders. Such an explanation proved satisfying for the farmers and they had no further complaints.

3.8. Conclusion: urban waste, new institution and combating desertification

From November 2005 to February 2018, we built 34 sites with a total area of 10.03 ha on which urban waste was used to make pastureland.[5] After receiving requests from Fulbe and Tuareg herders around the research village, I

---

5   In order to expand activities for land restoration, the author received support from the Mitsui & Co. Environment Fund for three years. I would like to show appreciation for their support.

arranged meetings with a herder client, his family and relatives, the village headman and influential persons of the Hausa people in order to discuss the possibility of site construction. They decided the placement of sites and took me to the planned sites. Then I worked at the site with the villagers, including farmers and herders, males and females, the young and the aged, persons with disabilities and so on. The city mayor issued a letter of permission every year for our activities. I gave the photocopied letters to both of the client herders and the village headmen.

My trial was originally based on the local ecological knowledge of Hausa farmers, carrying urban waste to degraded farmland as part of their everyday life. Around 2005, this knowledge began to expand to other towns in southern Niger. Zarma famers in Niamey now also use urban waste to fertilize their farmland. As I noticed the potential to reduce livestock-induced crop damage and armed conflicts, I began to experiment with the technique and to apply this knowledge in the social context. The beneficiaries are not only herders but also farmers, because the herders take care of the farmers' livestock and the farmers can fatten their livestock on the pastureland created. Moreover, both can avoid paying crop compensation for livestock-induced crop damage. This trial implies a win-win relationship, or at least not a complete win-lose situation - a positive fit to the social-environmental system, indicated by Haller et al. (2013).

I consider my trials, based on local ecological knowledge, as reverse thinking. From the perspective of outsiders, when considering the rural area of the Sahel, we tend to focus on the negative effects of plastic bags and vinyl products, but I was more concerned with the positive effects of combating desertification. Firstly, plastic bags can prevent soil moisture evaporation from the ground under conditions of strong sunshine and high temperature. Secondly, they can catch windblown sand and organic matter and retain it on the degraded land. Thirdly, the plastic bags create a suitable environment for termite colonies. Termites are regarded as harmful insects that damage wooden houses, but they create suitable physical conditions in soil for plant growth, as mentioned by Holt and Lepage (2000). These effects are promoted by the presence of plastic bags.

It is easy for outsiders living in developed countries to criticize the ignorance and mindlessness of farmers with regard to food and environmental safety by applying urban waste to their farm and pastureland. If I had not applied the urban waste to the degraded land in order to create new pastureland, I would not have understood why numerous farmers would continue to collect urban waste from towns and apply it to their farmland. For the farmers living in southern Niger, the urban waste is an important resource for improving soil fertility and restoring their farmland. It is used to prevent land degradation. If we tell them to stop this practice, they will reply, "How can we maintain soil fertility in the face of land degradation?" and "How can we harvest enough millet to sustain our families?"

This practice has to be considered against the large-scale initiatives of the Great Green Wall for the Sahara and the Sahel Initiative, with high investments and overall rather limited effects. It is a small-scale practice that restores degraded land and enables more understanding and the building of sustainable relations between the multi-ethnic and diverse resource users.

Currently, the prevention of desertification, poverty, hunger, armed conflict and terrorism is a serious problem in the Sahel, and there are trade-offs to be considered when highlighting environmental concerns related to the use of urban waste. In recent years, the international community has given much attention to hunger, poverty and terrorism in the Sahel because it is strongly linked to immigration problems in European countries. It is extremely difficult to solve all the problems in the region, but it is my opinion that military operations will not be sufficient to prevent armed conflicts and terrorism. Hunger and poverty are the drivers of armed conflicts and terrorism in the Sahel and, in turn, armed conflicts and terrorism further exacerbate hunger and poverty. Both natural and anthropogenic factors have increasingly combined to generate conflicts over the available resources in the Sahel.

Under these severe conditions, a deadlock has been reached, preventing the situation from being resolved. There is no room for afforestation in the farmland and many afforestation projects have attempted to chase the herders and their livestock away to protect tree seedlings. By combining scientific and local ecological knowledge, we can aim to combat desertification, hunger and poverty, and prevent armed conflicts and terrorism, with the cooperation of the residents, including urban dwellers, rural farmers and herders. Reverse thinking and counter-intuitive approaches combining local and scientific knowledge could provide mitigation effects which create room for potentially better and more sustainable solutions.

Acknowledgements

The author would like to express deep appreciation to Tobias Haller and Claudia Zingerli for offering insightful and stimulating comments on this paper.

This research was carried out with financial support from the Japan Society for the Promotion of Science, Grant-in-Aid for Scientific Research (KAKENHI 17H04506 and 17H02235)

3.9. References

Adamu, M. (1978). The Hausa factor in West African history. Ibadan: Ahmadu Bello University Press.

Adepegba, D. and Adegoke, E. A. (1974). A study of the compressive strength and stabilizing chemicals of termite mounds in Nigeria. Soil Science, 117(3), 175–179.

Ayantunde, A. A., Williams, T. O., Udo, H. M. J., Fernandez-Rivera, S. Hiernaux, P. and van Keulen, H. (2000). Herders' perception, practice, and problems of night grazing in the Sahel: Case studies from Niger. Human Ecology, 28(1), 109–130.

Bagine, R. K. N. (1984). Soil translocation by termites of the genus Odontotemes (Holmgran) (Isoptera: Macrotermitinae) in an arid area of northern Kenya. Oecologia, 64, 263–266.

Baier, S. (1980). An economic history of central Niger: Oxford studies in African affairs. Oxford: Clarendon Press.

Benemann, J. R. (1973). Nitrogen fixation in termites. Science, 181(4095), 164–165.

Bleich, K. E. and Hammer, R. (1996). Soils of Western Niger. In: Buerkert, B., Allison, B. E. and von Oppen, M. (eds.). Wind erosion in Niger: Implications and control measures in a millet based farming system (pp. 23–32). Dordrecht: Kluwer Academic Publishers.

Brouwer, J. and Powell, J. M. (1998). Increasing nutrient use efficiency in West African agriculture: The impact of micro-topography on nutrient leaching from cattle and sheep manure. Agriculture, Ecosystems and Environment, 71, 229–239.

Crang, M. and Cook, I. (2007). Doing ethnographies. Los Angeles: Sage Publications.

Dregne, H. E. (1986). Desertification of arid lands. In: El-Baz, F. and Hassan, H. A. (eds.). Physics of desertification (pp. 4–34). Dordrecht: Springer.

Fairhead, J. and Leach, M. (1995). False forest history, complicit social analysis: Rethinking some West African environmental narratives. World Development, 23(6), 1023–1035.

Gonzalez, P. (2001). Desertification and a shift of forest species in the West African Sahel. Climate Research, 17, 217–228.

Graef, F. and Haigis, J. (2001). Spatial and temporal rainfall variability in the Sahel and its effects on farmer's management strategies. Journal of Arid Environments, 48, 221–231.

Haller, T. (2002). Common property resource management, institutional change and conflicts in African floodplain wetland. The African Anthropologist, 9(1), 25–35.

Haller, T. (2003). Rules which pay are going to stay: Indigenous institutions, sustainable resource use and land tenure among the Ouldeme and Platha, Mandara Mountains, Northern Cameroon. In: Le Meur, P-Y. and Lund, C. (eds.). Everyday governance of land in Africa (pp. 117–134). APAD-Bulletin No 22. London: Lit Verlag.

Haller, T., Fokou, G., Mbeyale, G. and Meroka, P. (2013). How fit turns into misfit and back: Institutional transformations of pastoral commons in African floodplains. Ecology and Society, 18(1), 1–16.

Haller, T., van Dijk, Bollig, M., Greiner, C., Schareika, N. and Gabbert, C. (2016). Conflicts, security and marginalisation: Institutional change of the pastoral commons in a "glocal" world. Revue Scientifique et Technique (International Office of Epizootics), 35(2), 405–416

Heiss, J. P. (2015). Musa: An essay (or experiment) in the anthropology of the individual. Berlin: Duncker and Humbolt.

Hill, P. (1972). Rural Hausa: A village and a setting. Cambridge, UK: Cambridge University Press.

Holt, J. A. and Lepage, M. (2000). Termites and soil properties. In: Abe, Y., Bignell, D. E. and Higashi, T. (eds.). Termites: Evolution, sociality, symbioses, ecology (pp. 389–407). Dordrecht: Kluwer Academic Publishers.

IEP (Institute of Economics and Peace) (2015). Global Terrorism Index 2015. Measuring and understanding the impact of terrorism. IEP Report 36. Sydney, News York, Mexico City: IEP

Lee, K. E. and Wood, T. G. (1971). Termites and soils. London: Academic Press.

Leisinger, K. M., Schmitt, K. and ISNAR. (1995). Survival in the Sahel. The Hague: International Service for National Agricultural Research.

Lund, C. (1998). Law, power and politics in Niger: Land struggles and the rural code. Hamburg: Lit Verlag.

Michels, K., Sivakumar, M. V. K. and Allison, B. E. (1995). Wind erosion control using crop residue II: Effect on millet establishment and yields. Field Crops Research, 40, 111–118.

Moritz, M. (2006). The politics of permanent conflict: Farmer-herder conflicts in northern Cameroon. Canadian Journal of African Studies, 40, 101–126.

Mortimore, M. and Turner, B. (2005). Does the Sahelian smallholder's management of woodland, farm trees, rangeland support the hypothesis of human-induced desertification? Journal of Arid Environments, 63, 567–595.

Oyama, S. (2009). Ecological knowledge of Hausa cultivators for the land degradation process in Sahel, West Africa. Geographical Reports of Tokyo Metropolitan University, 44, 103–112.

Oyama, S. (2012). Land rehabilitation methods based on the refuse input: Local practices of Hausa farmers and application of indigenous knowledge in the Sahelian Niger. Pedologist, 55(3), 466–489.

Oyama, S. (2014). Farmer-herder conflicts, land rehabilitation, and conflict prevention in Sahel region of West Africa. African Study Monographs, 50(supplementary), 103–122.

Oyama, S. (2015a). Land degradation and ecological knowledge-based land rehabilitation: Hausa farmers and Fulbe herders in the Sahel region, West Africa. In: Reuter, T. (ed.). Averting a global environmental collapse: The role of anthropology and local knowledge (pp. 165–185). Cambridge, UK: Cambridge Scholars Publishing.

Oyama, S. (2015b). Tackling the land degradation in Sahel region of West Africa: Trash input for land rehabilitation, food security and conflict prevention. Kyoto: Showado (in Japanese).

Oyama, S. (2017). Hunger, poverty and economic differentiation generated by traditional custom in villages in the Sahel, West Africa. Japanese Journal of Human Geography, 69(1), 27–42.

Oyama, S. and Mammane, I. (2010). Ecological knowledge of Hausa cultivators and in situ experiment of the land rehabilitation in Sahel, West Africa. Geographical Repots of Tokyo Metropolitan University, 45, 31–43.

Pomeroy, D. E. (1976). Some effects of mound-building termites on soils in Uganda. Journal of Soil Science, 27, 377–394.

Prudat, B., Bloemertz, L. and Kuhn, N. (2018). Local soil quality assessment of north-central Namibia: Integrating farmers' and technical knowledge. Soil, 4, 47–62.

Richards, P. (1985). Indigenous agricultural revolution: Ecology and food production in West Africa. London: Unwin Hyman.

Schlecht, E. and Buerkert, A. (2004). Organic inputs and farmers' management strategies in millet fields of western Niger. Geoderma, 121, 271–289.

Shinjo, H., Hayashi, K., Abdoulaye, T. and Kosaki, T. (2008). Management of livestock excreta through corralling practice by sedentary pastoralists in the Sahelian region of West Africa: A case study in southwestern Niger. Tropical Agriculture and Development, 52(4), 97–103.

Suzuki, L., Matsunaga, R., Hayashi, K., Matsumoto, N., Tabo. R., Tobita, S. and Okada, K. (2014). Effects of traditional soil management practices on the nutrient status in Sahelian sandy soils of Niger, West Africa. Geoderma, 223–225, 1–8.

Tschakert, P. (2007). Views from the vulnerable: Understanding climatic and other stressors in the Sahel. Global Environmental Change, 17, 381–396.

Turner, M. D., Ayantunde, A. Patterson, K. P. and Patterson, E. D. (2011). Livelihood transitions and the changing nature of farmer-herder conflict in Sahelian West Africa. Journal of Development Studies, 47(2), 183–206.

4. Energy and the environment in Sub-Saharan Africa
Household perceptions of improved cookstoves

Sarah Jewitt, Peter Atagher, Mike Clifford, Charlotte Ray and Temilade Sesan

4.1 Introduction

Globally, 2.7 billion people rely on solid biomass fuels like fuelwood, charcoal, animal dung, grass, shrubs or agricultural residue to meet their cooking and heating needs. In Sub-Saharan Africa, 80% of households rely on biomass and many people cook on open fires inside their homes. As well as being fuel inefficient, household members - especially women and children - are exposed to harmful levels of wood smoke (Bruce et al. 2000, 2015). Analysis of data from the 2010 Global Burden of Disease study (Lim et al. 2012) identified household air pollution (HAP) arising from the burning of solid biomass fuels as the second most common cause of death in eastern, central and western Sub-Saharan Africa. The same study identified HAP as the third most important risk factor globally (second for women), causing an estimated 3,478,773 deaths annually and contributing 4.3% of global disability adjusted life years (DALYs). They key causes of death linked to HAP include chronic respiratory diseases, heart disease, childhood pneumonia, cancers, cataracts, and burns (GACC, nd, IEA and World Bank 2015).

4.1.1. The evolution of improved cookstove initiatives

Improved cookstoves (ICS) designed to burn biomass fuels more cleanly and efficiently have been promoted by charities, governments and private sector actors in low-income countries since the late 1940s. Some of the earliest designs sought to reduce smoke (Smith 1989, Sesan 2014) but by the early 1970s, concerns about fuelwood shortages causing deforestation (Eckholm 1975) helped to focus attention on more fuel-efficient stoves (Barnes et al. 1994). At the same time, attention focused on the gender implications of a reliance on biomass for cooking and heating as gender divisions of labour often give the responsibility for biomass collection to women and girls. This can involve them in spending many hours per day undertaking such work which can in turn compromise their access to education and the development of skills needed for a variety of income generating activities (Agarwal 1986). Following critiques of the scale of fuelwood-related deforestation, however, attention gradually re-focused on the health and gender impacts of cooking with biomass (Hanbar and Karve 2002, Nagothu 2001, Sesan 2014). More recently, concern has grown about the climate change impacts of traditional cookstoves given that they are estimated to emit a third of global carbon monoxide along with significant emissions of black carbon (soot), nitrous oxide, methane and non-methane volatile organic compounds (Rosenthal 2009, Venkataraman et al. 2010). This helped to consolidate a shift towards emphases on the use of "clean" fuels and cookstoves with potential to offer global environmental and health benefits (Hanbar and Karve 2002, Nagothu 2001, Simon 2010, Sesan 2014).

To an extent, this reflects the underlying assumptions of the "energy ladder" model that household fuel preferences shift, with increasing income, from a reliance on biomass fuel to transitional fuels such as kerosene and later to cleaner and more efficient fuels such as gas or electricity (Bruce et al. 2000). Nevertheless, a number of empirical studies have questioned the model, highlighting the low priority given to the adoption of modern fuels given competing household economic priorities and the ways in which households often combine (or "stack") different fuels in different seasons in order to undertake different types of cooking (Masera et al. 2000, van der Kroon et al. 2013, Ruiz-Mercado and Masera 2015, Treiber et al. 2015).

4.1.2. Recent initiatives promoting clean fuels and cookstoves

Despite a plethora of interventions, however, the adoption and sustained use of clean and improved cookstoves remained low globally, prompting the establishment, in 2010, of the Global Alliance for Clean Cookstoves (GACC) which seeks to "foster the adoption of clean cookstoves and fuels in 100 million households" by 2020 (GACC, 2017a). To this end, GACC promotes the use of the International Working Agreement's (IWA) stove tier system which sets out guidelines for rating stoves according to their efficiency/fuel use, safety, indoor emissions and total emissions. As with the energy ladder, there is an assumption that increasing socio-economic status encourages households to make a linear transition to the use of higher tier stoves (GACC 2017b; Ray et al. 2017)

Working alongside GACC, Sustainable Energy for All (SE4ALL) has focused on including energy in the post-2015 global goals and making efforts to meet the UN General Assembly's key targets to "ensure universal access to modern energy services, double the global rate of improvement in energy efficiency and double the share of renewable energy in the global energy mix" (IEA and World Bank 2015: 38). As part of this role, SE4ALL has collaborated with the Energy Sector Management Assistance Program (ESMAP), IEA and World Bank to prepare indicators for Sustainable Development Goal 7 (SDG7) which seeks to "ensure access to affordable, reliable, sustainable and modern energy for all" by 2030 (UN, 2016a). Echoing historical variations in the thrust of earlier cookstove interventions, SE4ALL and GACC have respectively emphasized how "non-solid fuels"[1] and clean cookstoves can provide health gains, improve fuel use efficiency and wider environmental benefits associated with lower levels of greenhouse gas emissions and forest decline (Lewis and Pattanayak 2012, Bielecki and Wingenbach 2014).[2] Recognizing the gender implications of a reliance on biomass fuels, GACC has also linked low levels of female empowerment to the failure to include energy in the Millennium Development Goals (GACC, nd).

SE4ALL, meanwhile, has been instrumental in developing a "global tracking framework" (GTF) to provide baseline data for SDG7 targets in terms of access to "modern cooking solutions" (IEA and World Bank 2015:

---

1   Solid fuel use (e.g. wood, crop residue, dung or charcoal) in low-income countries has been linked to inefficient combustion and negative health impacts while the use of non-solid fuels such as biogas, LPG, electricity, ethanol, natural gas and solar energy (BLEENS) is associated with more efficient and cleaner, healthier cooking practices (IEA and World Bank, 2015). Although kerosene is classed as a non-solid fuel, it tends to be excluded from this group of more desirable fuels because of the pollution it causes as well as the risk it presents in terms of burn-related domestic injuries.

2   GACC's target of clean cookstove adoption by 100 million households (of the estimated 2.9 billion that rely primarily on solid fuels – GTF 2015), echoes the target of Millennium Development Goal (MDG) 7C to halve the proportion of the population without sustainable access to basic sanitation which was criticised for focusing on promoting uptake among the "low hanging fruit" of higher income with no previous sanitation access.

48) and the "percentage of population with primary reliance on non-solid fuels" (IEA and World Bank 2015: 3). These data suggest that the number of solid fuel users globally rose from 2.8-2.9 billion with significant inequalities in access to non-solid fuels (5% and 40% respectively) between poorer and wealthier groups (IEA and World Bank 2015: 55-56). Unlike the GTF, however, SDG indicator 7.1.2 which tracks the "proportion of population with primary reliance on clean fuels and technology" (UN 2016b) makes no reference to either non-solid fuels or cooking. SDG7's retention of the GTF's focus on tracking "primary reliance" on cooking fuels and technologies, meanwhile, discourages data collection on whether households use of a range of different fuels (stacking) in order to adjust to factors like fluctuating fuel prices, seasonal fuel availability or changes in the number of people they are cooking for (Masera et al. 2000, Ruiz-Mercado et al. 2011, Rehfuess et al. 2014, Ruiz-Mercado and Masera 2015, Loo et al. 2016, Lozier et al. 2016).

4.1.3. Neglect of end-user preferences

Despite emphases by GACC and SE4ALL on the environmental, health and gender benefits associated with clean fuel and cookstove use, end-user perspectives continue to be marginalized and there has been limited evidence to date of the use of participatory approaches to either better understand barriers to the adoption of clean fuel and cookstoves or promote their use. According to Sesan (2014: 6) initiatives promoting more efficient cookstoves in the 1970s and 1980s were characterized as a "straightforward technical challenge" with limited end-user engagement. From the 1990s, emphasis within the stove sector shifted towards more commercially-oriented initiatives producing efficient but often unaffordable stoves lacking key features prioritized by end-users (Simon 2010, Sesan 2014, Jewitt and Rahman 2017).

A common feature of improved cookstove initiatives has been the use of more quantitative, techno-centric approaches that produce highly efficient stoves that lack key functions required or desired by their end-users (IEEE 2014, Ray et al. 2014, 2017). This has occurred despite the weak relationship between socio-economic status and the use of biomass for cooking pointing to low levels of demand for "clean" or "efficient" stoves and strong user preferences for solid fuels (IEA and World Bank 2015: 63). After all, in areas where solid fuel can be gathered free of cost, stoves that require fuel to be purchased are unlikely to be attractive given other demands on household budgets.

Efforts to address similar constraints in the sanitation sector ("free of cost" open defecation versus costly sanitation systems), also met with frequent failure when technology-oriented initiatives were employed in areas with low demand for latrines. This prompted the development of social marketing approaches, participatory community-led "total sanitation" initiatives (CLTS) that proved far more successful in stimulating sanitation uptake (Kar 2003, Evans 2005, Kar and Pasteur 2005, Jenkins and Sugden 2006, Jenkins and Scott 2007, Peal et al. 2010, O'Reilly and Louis 2014, Arickal and Khanna 2015).

Common examples of participatory tools used in CLTS include social mapping exercises to identify commonly used open defection (OD) sites and transect walks to illustrate faecal-oral transmission routes between OD and food preparation sites. These are supported by community-designed mechanisms such as regular monitoring of OD sites to create social pressure to maintain this behaviour. This is important in helping to promote the realization that individual wider environmental health benefits can only occur if change occurs at the community level (Kar 2003, Kar and Pasteur 2005). Until quite recently, however, evidence of such approaches being used for cookstoves has been limited (Graham 2015, Rosenbaum 2015).

4.1.4. Limitations of fuel and ICS monitoring

Compounding the drawbacks associated with poor end-user engagement in cookstove promotion initiatives is a tendency for existing fuel and cookstove use monitoring mechanisms to ignore the complexity and fluidity of household energy use. In contrast to the energy ladder's assumptions of upwards progress in a linear manner, families are just as likely to move "down" the ladder in response to rising prices for their primary cooking fuel (IEA and the World Bank 2015). Their "primary" fuel use may also vary seasonally in response to changes in weather or resource availability while the importance of fuel or stove "stacking" (Masera et al. 2000) is unlikely to be captured using current monitoring approaches.

Reflecting these shortcomings, improved cookstove (ICS) interventions and monitoring systems have been criticized for failing to understand the complex ways in which household cooking systems are embedded in local cultures and livelihoods (Ruiz-Mercado et al. 2011, Ray et al. 2014, Sesan 2014, Ruiz-Mercado and Masera 2015). At the same time, there is increasing recognition within the ICS literature of the need to understand and respond to the priorities and preferences of end-users (Beyene and Koch 2013, Kohlin et al. 2011). Increasingly, research has called for the greater use of qualitative, participatory methods to explore the non-technical dimensions of ICS dissemination and understand socio-economic and cultural factors affecting fuel and stove choices (Ray et al. 2014, 2017).

4.1.5. Research problem and contribution

Drawing on participatory "bake/cook-off" events in the UK, Malawi and Zambia plus empirical evidence from Benue State, Nigeria, sections 3 and 4 of this chapter provide insights into how context-specific end-user priorities coupled with constraints associated with different settings often inhibit a linear shift towards sustained use of one clean cooking system. Attention is drawn to how cooking practices, preferences and taboos tend to vary over space with factors such as socio-economic status, environmental change, cultural norms associated with cooking, fuel availability/cost, family size, ethnicity, age or gender often having a significant influence on both household energy preferences and whether a particular fuel or cookstove is likely to be accepted and adopted. Particular emphasis is placed on how households understand and access environmental resources in order to meet their daily energy needs and why many still prefer to use solid biomass for their cooking needs; especially where it can be gathered free of cost.

The chapter's originality and rigour lies in its use of qualitative methods along with participatory approaches to obtain end-user priorities for cooking fuels and technologies in contexts where households commonly use a range of different systems and make frequent shifts between them. Its significance lies in its emphasis on the need to develop participatory approaches that will help to improve monitoring and better understand end-user preferences and engage them in ICS design, production and dissemination initiatives.

4.2. Methodological approaches

A range of qualitative methods and participatory approaches were utilized in this research as a means of developing innovative approaches for sharing interdisciplinary academic and user-based perspectives on "improved" cookstove and household energy systems. Phase one was associated with a series of "bake/cook-off" events to elicit end-user perspectives on a range of ICS while phase 2 involved in-depth field-based research on cooking practices and priorities in Benue state, Nigeria (see Figure 1).

Figure 1: Map of bake/cook-off events and field-based research

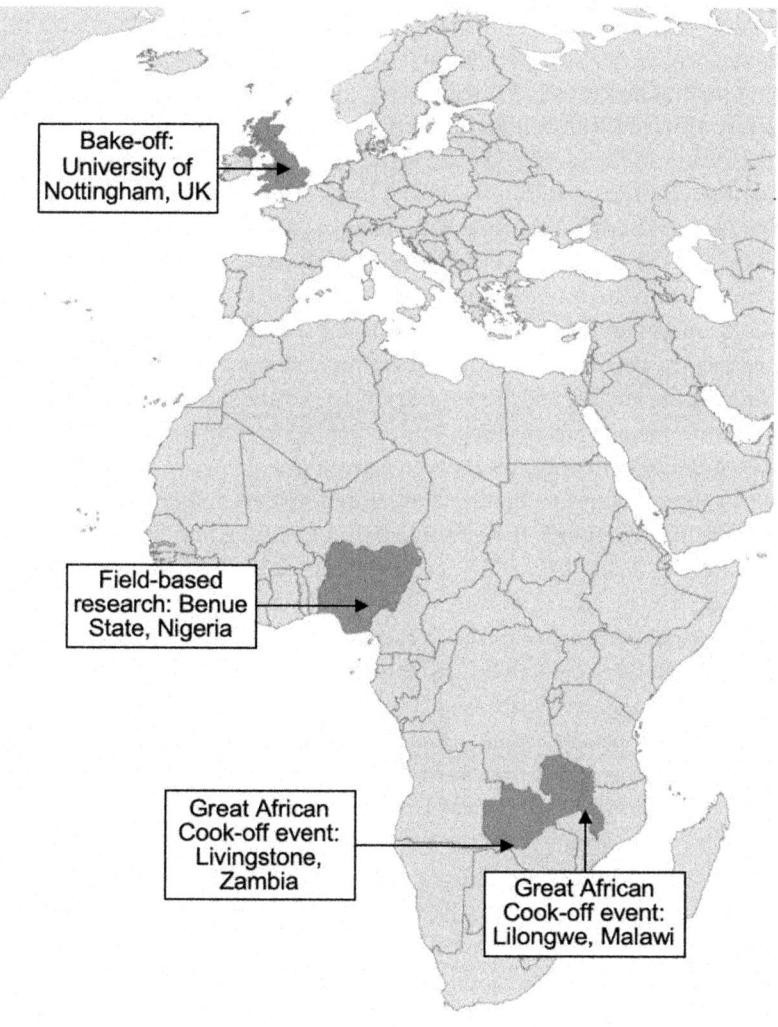

4.2.1. Bake/cook-off events

The bake-off event that took place at Nottingham University in September 2015 was attended by volunteers from the Nottingham Women's Cultural Exchange, academic researchers, development practitioners and policy makers. The volunteers came from a range of African countries (including Nigeria, Eritrea, Sudan, and Malawi) and all had previous experience of cooking with biomass fuels. At the event, a range of improved cookstoves were made available and the volunteers were invited to cook typical food from their home countries on their chosen stove and share their experiences of using this stove with the workshop attendees. A key aim of the bake-off was to create opportunities to observe different ICS in action and how end-users interact with them as a means promoting discussion and enhancing understandings of user preferences, performance, safety and wider cultural (especially gender) considerations surrounding energy/fuel choice.

To encourage broader discussions about the different technologies used at the event, a range of participatory exercises were used to help volunteers and attendees to identify and discuss what they liked and disliked about the different stoves and compare views. This in turn led to fruitful discussions about how differences in priorities between policy makers, stove manufacturers and end-users could impact on the adoption and sustained use of ICS.

The "bake/cook-off" format was adapted for use in Malawi in March 2016 as part of a "Great African Cook-off" event at the Cleaner Cooking Camp Conference[3] which brought together national and international stove enthusiasts to discuss challenges around clean cooking in Malawi and test a range of stoves for end-user acceptability. As with the Nottingham bake-off, this event provided a designated space for attendees to interact with participating cooks and was attended by a range of stakeholders including government, donor organizations and INGOs as well as the Malawian Minister for Energy. Later in 2016, a second "Great African Cook-off" event took place in Livingstone, Zambia, where members of the public, stove producers, policy makers and charitable organizations gathered to share knowledge and learn about rocket stoves, solar cookers, imported gasifier stoves, improved charcoal burners and handmade clay rocket stoves. As insights gained from the Nottingham event played a key role in shaping the research questions and methodologies used

---

[3] This annual event is supported by the National Cookstoves Steering Committee and led by the Energizing Development Programme (EnDev).

in the field-based research in Benue, these will be discussed in more detail below.

4.2.2. Field-based research in Benue State

Benue state was chosen as a location for the field-based research on the basis of the high but declining level of dependence on wood as a domestic cooking fuel coupled with one of the author's familiarity with the area and his ability to speak Tiv. From the perspective of undertaking participatory research, this author's role as an "insider" as well as an "outsider" (born and raised in Benue state but educated overseas and undertaking research in Benue) placed him in an excellent position to mediate between "outside" researchers and local community members (Mosse 2008) whilst translating the research agenda into a methodological approach that engaged participants. At the same time, his familiarity with local socio-economic and cultural norms helped in gaining access to local community members and building trust; eliciting, in the process, information regarding changes in local cooking practices and preferences and the cultural norms surrounding them. Local knowledge of wider policies affecting fuel and stove availability, seasonal weather patterns, their influence on employment opportunities and associated rural-urban migration rhythms were also important for planning appropriate times for undertaking different elements of the research. The "outsider" perspectives of other team members and bake/cook-off participants, meanwhile, helped to ensure that research questions and methodological approaches arising from these events along with shortcomings associated with broader fuel and cookstove dissemination and monitoring trends could be addressed by the methodologies employed in the study sites.

Benue is located in north-central Nigeria and has an estimated population of 4 million in an area of 30,800 km^2. The dominant ethnic groups in Benue are the Tiv with around 69% of the population followed by the Idoma and the Igede which make up around 23% of the population (NBS/CBN/NCC 2011). According to Ali and Victor (2013), socio-economic development in the state is strongly dependent on the charcoal and firewood trades. Dapo and Emmanuel (2013) found that the majority (76.7%) of households that used charcoal as a cooking fuel spent an average of N3310[4] on this monthly, while 23.3 percent of households spent N2394 monthly on alternative cooking fuels. Firewood

---

4 At the time of the research, £1 was N396

and charcoal are more readily available in the state than kerosene so there are competing demands for these fuels for cooking and other exigencies. Both fuels can be purchased from roadside traders although many households are willing to travel several kilometres to collect firewood for free.

In order to provide to help with the selection of sites for community-based data collection, state-level data from the 2008, 2010, 2013 and 2015 Demographic Health Surveys (DHS) for Nigeria were obtained and information on the type of fuels household mainly use for cooking was analyzed for rural and urban areas of Benue using SPSS. This revealed that in 2008, 93.4% of households used wood as their main cooking fuel with 4.5% using kerosene and 0.3% using charcoal. A very small number of households used natural gas but none reported electricity or Liquefied Petroleum Gas (LPG) as their main cooking fuel in the 2008 survey. Subsequent surveys revealed a slight decline in the use of wood as the main cooking fuel (92% in 2010, 88.2% in 2013 and 79.1% in 2015) while kerosene use rose slightly before declining (6.5% in 2010, 8.5% in 2013 and 6.6% in 2015) and charcoal use increased (2% in 2013 and 6.2% in 2015) along with natural gas (negligible in 2010 and 1.4% in 2015)

DHS data also revealed significant variations between rural and urban areas, however, with 42% of urban households relying on wood, 44% relying on kerosene, 7% on electricity, 5% on charcoal and 2% on LPG as their main cooking fuel compared to 84% relying on wood, 10% on kerosene, 5% on electricity and 1% on charcoal in rural areas. According to surveys conducted by the National Bureau of Statistic (NBS), Central Bank of Nigeria (CBN) and National Communication Commission (NCC) conducted across Nigeria in 2010, urban households spend an average of N970 and N1233 per month on firewood and charcoal (NBS/CBN/NCC 2011).

Information was obtained from three sites chosen on the basis of their contrasting fuel availability. Site 1 is an urban community within the state capital, Makurdi, where LPG, electricity and charcoal are available in addition to fuelwood. LPG is available from a gas refilling plant where a 10kg cylinder can be refilled for N1800. The cost of gas stoves ranges from N12672 to N93000 while a two burner electric stove costs N21,000. Demand for firewood here has increased as a result of brick-making activities in the area which has forced up prices. The town's proximity to the state capital coupled with the presence of a gas refilling plant has enabled some households to access to ICS and modern fuels such as electricity and gas thus providing useful insights into the value attached to ICS and modern fuels compared to more "traditional" cooking systems. The presence of a brewing company has encouraged in-migration for employment purpose which has in turn stimulated the development of a range of food restaurants which - in the absence of large capacity improved cookstoves - cater for the town's population using wood on open fires as their key fuel. Restaurant owners along with many local residents travel several miles outside the community to collect free firewood.

Site 2 is a peri-urban community with a long-standing reliance on firewood as the primary cooking fuel. Firewood is purchased from communities across the river and transported by canoes. Households that do not have the economic resources to purchase firewood travel several kilometers outside the community to gather it free of cost as local woodland areas have been exhausted. Most households prepare meals on three-stone fires placed outside their dwellings although some make additional use of kerosene and "Abacha" stoves which are usually constructed of steel and use charcoal as their primary source cooking fuel.[5] Kerosene stoves cost around N8000 compared to N3000 for an Abacha stove.

Site 3 is a rural community and households travel shorter distances to obtain their cooking fuel as they have access to state-managed plantations. Livelihoods are dominated by trading and the processing of farm produce although fuelwood selling is also widespread. Although some households own gas, kerosene and Abacha stoves, many have reverted to using three stone fires to reduce the costs of purchasing fuel. Restaurant owners here cook on three-stone fires rather than improved stoves as they consider the former the best option for cooking large quantities of food.

4.2.3. Field-based methodologies

The research was conducted in two phases with findings from the pilot phase being used to refine the research questions. The targets of the community-based research were households from different socio-economic and ethnic groups in the three study sites that used a range of different fuel and cookstove types. The study employed household surveys, focus group discussions, participatory appraisal tools and direct observation to understand decisions and preferences relating to cookstove and fuel use in the context of broader household socio-economic priorities. Participants that took part in focus group discussions during the pilot phase were later re-visited to enable more

---

5   Abacha stoves were introduced in Nigeria in 1994 by the Military Head of State General Sani Abacha in response to kerosene shortages and resulting price hikes.

detailed information on household cooking practices and preferences to be obtained. In order to gather gendered perspectives on household energy and cooking preferences, data was collected by both male and female researchers.

Drawing on emphasis by Ruiz-Mercado and Masera (2015) on the need to understand the cultural dynamics driving households' fuel choices and cooking device priorities, focus group discussions and participatory exercises were carried out with 49 (14 male, 35 female) participants. At the same time, participatory ranking exercises were undertaken to provide understandings of where stove and fuel choice sat within broader household priorities and aspirations. In each study site, 21 household surveys were undertaken to obtain information on household demographics, existing fuel and stove use, fuel and cooking preferences and perceptions of different stove features. At the same time, household observations were carried out and fieldnotes were taken in order to obtain complementary data for comparison across the sample categories.

Semi-structured interviews were undertaken with 3 male community elders in each study site to elicit information on existing cooking practices and the extent to which these have changed over time. Two (male) stove artisans were also interviewed to elicit information on State government strategies and policy support programmes (if any) for developing the sector. Additional elite interviews were carried out with government employees in relevant Ministries. At the state level, the Director of Forestry in the Ministry of Environment and Urban Development, Benue State, was interviewed to explore programmes related to ICS interventions and energy policy in Benue. At the national level, interviews were conducted with a scientific officer within the Energy Commission and with a director in the Ministry of Women's Affairs and Social Development as both had responsibility for implementing energy-related policy decisions. Qualitative data derived from these methods were transcribed and exported into NVivo 10, coded and analyzed qualitatively, using thematic analysis to group emerging themes. Quantitative data from the household surveys were coded in Microsoft Excel 2013 and analyzed with a zero non-response.

4.3. End-user priorities for cooking systems: results from the bake/cook-off events

All three bake/cook-off events were highly participatory in nature as they focused around end-users identifying criteria that they associated with "improved" cookstoves and then choosing one or more ICS to prepare food on (see Figure 2).

Each event then provided opportunities to use a range of improved cookstoves with a key theme being the sharing of multi-disciplinary and user-oriented perspectives on what is understood as an "improved" cooking or household energy system. At the Nottingham bake-off the range of different stoves in use and on display and the food being cooked provided foci for the discussion and elicited a range of questions from participants including academics with different disciplinary backgrounds (engineering, science and technology studies, education, psychology, human factors, business, nursing, health sciences, sociology, development studies, geography, built environment), members of the public, development practitioners, NGOs and policy makers.

The fact that participants were able to experience key stove characteristics first hand (e.g. smokiness, controllability, stability, cooking speed, fuel efficiency etc.) was enormously effective in enhancing understandings of the advantages and constraints associated with using different stoves while discussions with cooks on their cooking experiences provided important insights into wider socio-cultural practices surrounding household energy use in energy-poor low-income country contexts.

At the end of the bake/cook-off sessions, participants were asked to note down their observations about the different stoves they viewed in operation as well as an additional selection of stoves that were left out for viewing but not lit. As a key emphasis was to collect views from a range of disciplinary and stakeholder perspectives, they were not provided with any guidance on how to structure their comments. Key observations focused around affordability, safety (especially linked to re-fuelling mechanisms but also stability-related issues), smokiness (especially in confined spaces), efficiency, durability, controllability and versatility (see Figure 3)

The cooks, meanwhile, were asked to re-visit the criteria they associated with ICS and then undertake a participatory ranking exercise to examine the performance of each stove according to these different criteria. As part of this process, they were encouraged to discuss wider cultural (including gender) considerations surrounding energy/fuel preferences and share experi-

Figure 2: Great African Cook-Off in Malawi. Picture: Charlotte Ray and Maria Beard

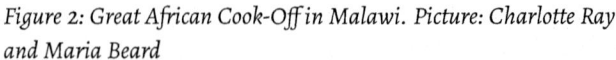

Figure 3: Word cloud of participant observations from the Nottingham bake-off

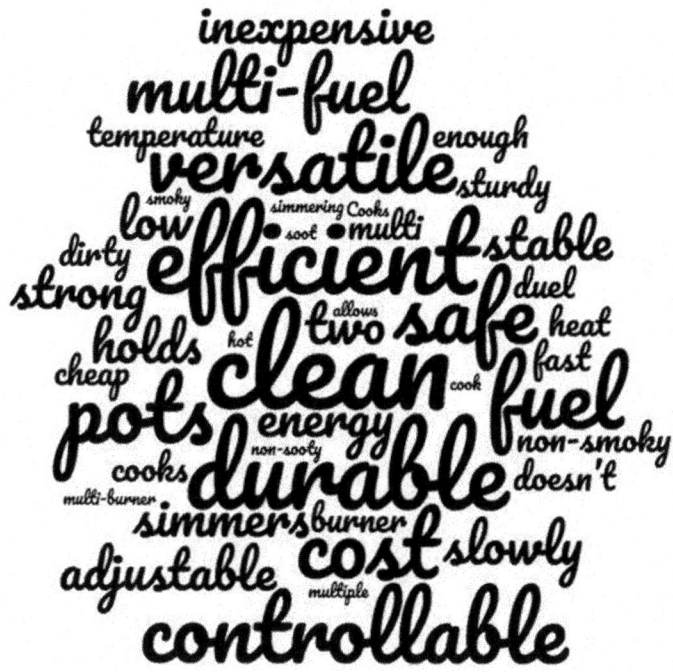

ences regarding variations in user priorities in different socio-economic and cultural settings. As can be seen from Figures 3 and 4, many of the criteria identified prior to the bake-off remained the same, but having used the different stoves, a number of additional criteria were identified for the matrix ranking exercise (see Figure 5). At the Nottingham bake-off, these focused mainly around safety issues including stability and burn-related risks for adults (linked to lighting and re-fuelling the stoves) and children. Criteria linked to controllability, versatility and cleanliness, meanwhile, increased in importance as a result of widespread admiration among the cooks for the EcoZoom La Plancha Stove with dual hotplates and an oven, the Ace 1 stove with a USB charger and the Clean cookstove which offered controllability and did not dirty the cooking pots. These user-preferences are reflected in the final matrix ranking in which the La Plancha and Ace stoves scored the highest on usability-related criteria by the cooks but were considered expensive (around £320 and £130 respectively) so received low scores for cost.

Figure 4: Word cloud of participant observations from the Nottingham bake-off

Discussions of stove "stacking" also took place as the cooks debated which stoves were best for cooking different types of food along with external influences (e.g. rules associated with living in rented accommodation) on the types of cooking systems that could be used. Other interesting observations that arose from discussions between cooks and with participants focused around how easy the different stoves were to light, how quickly they reached cooking temperature and the ease with which they could be re-fuelled. The latter

Figure 5: Cooks' matrix ranking from Nottingham bake-off

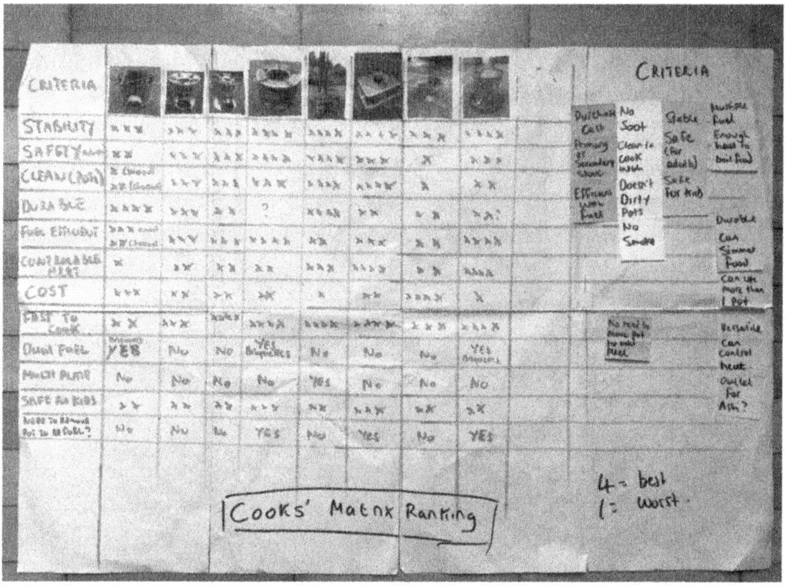

discussions were particularly interesting as they brought to light concerns associated with taking the pot off the stove to add fuel. Some cooks discussed the safety implications of removing a pot full of hot liquid to add fuel while others mentioned that removing the pot from the flame is regarded as taboo in some cultural contexts.

Accounts from different domestic settings, meanwhile, indicated that household members with responsibility for cooking, fuel gathering and cleaning dirty pots often had little control over domestic budgets so in the absence of serious health problems, changes would be unlikely to occur. In the Nottingham bake-off, where the cooks mostly had a background as refugees or asylum seekers and had become used to using electric or gas cookers in the UK, the smokiness and dirty pots produced by the charcoal and fire-wood fuelled stoves were a greater concern than it was the case in the Malawi and Zambia cook-off events where the cooks had less exposure to non-solid fuels. These observations along with those of the workshop participants were later transcribed and used to inform the content of the household surveys, semi-structured interviews, focus group discussions and participatory tools used in the Benue fieldwork.

4.4. Community-level perspectives on cooking systems and fuel choices in Benue

The main period of fieldwork in Benue took place over a period of six months between December 2015 and May 2016 although a pilot phase was undertaken in Spring 2015 with the purpose of making contact with the proposed study communities, piloting the household survey and undertaking some focus group discussions to identify key issues relating to fuel and cookstove use and preferences in the different sites.

4.4.1. Class and gender as influences on ICS and fuel use

Echoing GTF data indicating low uptake of non-solid fuel use by low-income groups (IEA and World Bank 2015), financial constraints were highlighted in the study sites as a key influence on cooking system choices, although broader cultural influences, such as Tiv traditions of hospitality, sometimes intersected these choices. Aside from a reliance on three stone fireplaces when catering for social gatherings or large family sizes, group discussions indicated that low-income groups tended to use three stone fireplaces with firewood while the use of clean cooking technologies such as gas and electric stoves was more pronounced amongst higher income groups.[6] As one low-income householder mentioned:

> "We have limited resources ... so we predominately cook with firewood" (Interview two: Site 2, 2015).

In addition to income status, gender has an important influence on decisions about stove and fuel choice in the study sites with men tending to defer to women in recognition of their responsibility for household food preparation. As one female respondent noted:

---

6   Income groups in the study sites were identified using a range of indictors including number of income-earners in the household, their approximate income levels and key household expenses including the average cost of household energy.

"My husband understands that I am the one who is always in the kitchen and would be in a better position to understand my needs so when I ask him to purchase anything in the kitchen including a stove he quickly obliged" (Focus group one: Site 3, 2015).

This view was also echoed by most male focus group participants who indicated that decisions of stove purchases would be largely in the hands of women; even at the point where requests for household finances were made. As two male respondents noted:

"A woman ... ordinarily she is the cook and our culture is such that she is the 'commander' in the kitchen so whatever she said at any moment is the last order and the husband has to obey, so the decision rests with the woman" (Focus group one: Site 1, 2015).

"Cooking duties are exclusively reserved for women and men are strongly forewarned to stay clear of cooking" (Interview two, Site 2, 2015).

One negative impact of women's responsibility for stove and fuel choices is that they also tend to have responsibility for collecting fuel for these stoves. This often involves significant drudgery and when men often get time to relax after returning from the fields, women often have to go in search of fuelwood and water to enable them to fulfil their domestic responsibilities. This is particularly burdensome for families that have to travel long distances to collect fuelwood.

4.4.2. Access to firewood

Firewood availability was mentioned as a particular constraint in Site 2 although interestingly, this was not linked to deforestation but rather to the conversion of former forest and bush land into farmland. A consequence of this was that household members now have to travel up to 6km to collect free fuel and there were many complaints of firewood collection having become very "time-consuming" and "a tedious process" (Focus group two: Site 2, 2015). Respondents reported that:

"Since firewood around the community has been exhausted we travel to the hinterland covering 10km each trip. Although each of the journeys is not pleasing, we live like a family now. When anyone brings it home ... it has to be shared among the households that don't have strength to cover such distances" (Female. Focus group one: Site 2, 2015).

> "Sometimes when I go out to collect firewood ... when I have a lot of visitors ... It's really difficult ... difficult to get the quantity that will be enough to cook a meal that will accommodate everybody, so we have to go back again and again. This is tedious and we are suffering so much because of firewood collection but we don't have a choice" (Female. Focus group one: Site 3, 2015).

Other households had felt compelled by time poverty to start purchasing their firewood and complaints about the cost of firewood were common with one respondent stating that:

> "Firewood ... in this community is our major problem, N100[7] worth of firewood won't be enough to prepare a meal for a family of three and firewood is almost going into extinction" (Male village elder. Interview one: Site 2, 2015).

4.4.3. Smoke-related concerns versus household budget constraints

Many respondents acknowledged how the discomfort of the smoke generated by three stone fireplaces and in some cases, concern about the health impacts of smoke had resulted in a desire to shift to a "cleaner" stove:

> "My wife ... was experiencing pains in her eyes and when I took her to the hospital, I ended up spending so much money such that I have no savings again. Since then if she makes a demand in the kitchen, I quickly respond to it ... unless I don't have [the means to do so]" (Male. Focus group two: Site 3, 2015).
>
> "When the smoke becomes so intense we make a demand that a stove be purchased. So women are the ones that make a demand that a stove be purchased" (Female. Focus group two: Site 2, 2015).

There was also some acknowledgement of the advantages of "cleaner" stoves in terms of cleaner homes or less drudgery associated with cleaning sooty pans:

> "Since I have been using the Abacha stove, my cooking pots have remained clean but the three-stone fire produces a lot of smoke and dirt ... you don't experience that with Abacha stoves" (Female, Focus group one. Site 1, 2015).
>
> "Three stone fires and firewood smoke makes the kitchen look untidy including the cooking pots" (Female village elder. Interview two: Site 3, 2015).

---

7   £0.33 in February 2015.

Other respondents acknowledged the nuisance caused by smoke but said that they felt financially constrained from moving to higher tier cooking devices:

> "I am not comfortable cooking with firewood, and since I don't have money to buy kerosene I use firewood on my three-stone fire" (Female. Focus group one: Site 3, 2015).
>
> "Smoke ... I am not comfortable with it ... when I cook with firewood on my three-stone fire ... it is difficult because of smoke ... and I have to close the kitchen door and stay outside ... periodically I go back inside the kitchen to tend the fires. It is not my desire to cook on a three-stone fire but in our community I have to cook in this way since I don't have alternatives" (Female. Focus group one: Site 2, 2015).
>
> "Despite our ... awareness of different cooking technologies ... we still use firewood because we have limited resources to purchase these technologies ... We are aware of gas stoves for example but as I said earlier we are constrained by our low levels of income so the majority of people in this community cannot afford to redirect it to other non-profit yielding ventures" (Male village elder. Interview two: Site 2, 2015).

As the above quote suggests, despite the difficulties and costs associated with obtaining firewood and the smokiness of three stone fires, purchasing improved cookstoves was not regarded as a priority although some respondents suggested that they might be more popular in urban areas where the cost of fuelwood is higher. Even amongst a group of higher income households, however, improved cooking devices were placed below children's education, owning a business and owning property in a list of household priorities.

4.4.4. Socio-cultural factors influencing stove and fuel stacking

Further probing on this point coupled with direct observation revealed that most households owned more than one type of cooking device and the use of two devices at the same time was quite common depending on user priorities and the type or quantity of food being cooked. Echoing the literature on stove stacking (Masera et al. 2000, Ruiz-Mercado et al. 2011, Rehfuess et al. 2014, Ruiz-Mercado and Masera 2015), this allowed households flexibility to switch between cooking devices according to particular user preferences, changes in fuel availability or costs and household cooking requirements. Among households that owned a range of cooking devices, it was clear that reliability and maintenance considerations were important influences on their willingness to purchase a new stove.

Some also raised concerns about the reliability of improved stoves and how easily they could be repaired if problems arose with them:

> "If we have good artisans that can repair the stove then we won't have any problems; otherwise it will be difficult to use and maintain the stove" (Female. Focus group two: Site 2, 2015).

As only site 1 had a stove repair workshop, however, a lack of access to stove repair artisans was identified as a barrier to the adoption of ICS that respondents were unfamiliar with or did not trust the quality and robustness of.

In contrast with the energy ladder model, households were found to shift down as well as up tiers in response to changing financial circumstances or fuel costs:

> "I cook mainly on my three-stone fire ... though I have electric, kerosene and Abacha stoves but ... [the] electric stove, has high electricity bills associated with it. I can't afford continuous usage and the price of kerosene too is high[8] but the Abacha stove is okay given my lean resources" (Female. Focus group one: Site 1, 2015).
>
> "I have a gas stove, though I stopped using it because of the refilling, transportation charges ... all these have been ... major problems and I now use my Abacha stove" (Female. Focus group two: Site 2, 2015).
>
> "I have a kerosene stove as well as my three-stone fire but I don't use it any longer since kerosene is very expensive. I now use my three-stone fire although the price of firewood is almost the same as kerosene" (Female. Focus group one: Site 2, 2015).
>
> "We cook mainly on the three-stone fire ... though we have electric, kerosene and Abacha stoves, my daughters prefer the three-stone fire to these stoves" (Male. Focus group one: Site 1, 2015).

Other respondents described how they would choose their cooking devices according to the social situation they found themselves in:

> "I have a kerosene stove and my three-stone fire and I use them at the same time to prepare meals" (Male. Focus group two: Site 3, 2015).

---

8   Kerosene is sold at N115 in most gas stations in the study area.

> "I use my gas cooker when I have visitors because I normally want to stay around them so I cook inside the kitchen which is close to my sitting room but when I am with my family I cook their meals on an Abacha stove" (Female. Focus group one: Site 1 2015).
>
> "I have two stoves (kerosene and Abacha) in addition to my three-stone fire. If needed we would use all of them at the same time to cook the quantity of meal that would satisfy all visitors" (Male. Focus group two: Site 3, 2015).

In Tiv households where it is a cultural tradition for households to prepare large quantities of food as an indicator of socio-economic status, three stone fireplaces were particularly favoured as they allowed the cooking of large meals and/or catering for larger family sizes. Some participants reported that as a large household size is viewed as a blessing, they feel the need to prepare large quantities of food on a regular basis to satisfy the household as well as visitors. These results suggest families may outgrow their smaller improved cookstoves and revert to three stone fireplaces that can accommodate a larger pot:

> "I usually have a lot of visitors and people living with me so [an improved] stove may not be able to cook the desired quantity of food needed to entertain my guests" (Male. Interview one, Site 2, 2015).
>
> "I have stopped using my kerosene stove since my family size is now large and I have gone back to my traditional three-stone fire since it can cook the desired quantity of meals at once" (Female. Focus group one: Site 1, 2015).

Kerosene or Abacha stoves were also commonly used in these households to cook smaller or quicker meals with one respondent reporting the use of "the three-stone fire for preparing large quantities of food while the kerosene stove is for soup only" (Female. Focus group two: Site 1, 2015). Versatility and controllability were also noted as desirable stove characteristics.

> "I desire a stove that I will regulate the amount of heat to my cooking pot at the same time accommodate large pot sizes" (Female village elder. Interview one: Site 2).

Cultural preferences for the food cooked in particular ways were also mentioned as an influence on cooking system choice with respondents highlighting the benefits of smoke for food preservation and taste:

> "On the three-stone fire we use firewood as the main fuel for cooking and sometimes for preserving meat, which is one of the underlying traditional

cooking practice. This is so because you cannot use another stove for meat preservation, so when it comes to that ... firewood and the three-stone fire is utilized to give the meat an accentuated aroma that is highly appreciated. If I had a gas stove, I would still use firewood to preserve my meat because it is very important to the family" (Male. Interview one: Site 1 2015).

4.4.5. User preferences for rapid cooking

As illustrated in the quotes above, the combination of three stone fireplaces and firewood were also favoured on the basis of perceptions that they cook food quickly and save time. Typical responses from focus group participants included claims that firewood "cooks faster than any other fuel". Further discussions on this topic revealed that user perceptions about cooking speed were linked to flexibility in the amount of fuel that could be used on them. One respondent explained that:

> "With my three-stone fire I put in as much firewood as I can to enable my meals to cook faster" (Male. Focus group one: Site 1 2015).

Further discussions on this topic reflected a desire by male household members for their meals to be prepared quickly when they returned home. This encourages cooks to add more wood to their three-stone fires thereby increasing the heat and helping to perpetuate the belief that these traditional stoves cook faster than other devices.

4.4.6. Seasonal shifts in stove and fuel use

Nevertheless, the use of three-stone fireplaces tends to vary seasonally as they are often located outside the home and it is difficult to relocate them indoors during the rainy season. An additional problem is that households that collect firewood free of cost are forced to cook with wet wood at this time of year causing greater smoke emissions during cooking. Even firewood vendors struggle to keep their wood dry during the rainy season. Many villagers associated this with increased health problems and in addition to the male respondent who reported taking his wife to hospital with eye pain, a female respondent from site 3 recounted having spent a significant portion of her savings when her child was hospitalized in the 2014 rainy season as a result of smoke from wet firewood. During focus group discussions, the topic of increased smoke from cooking with wet firewood was a common theme with some respondents noting a desire to purchase cleaner stove/fuel combinations:

> "During rainy seasons, women experience difficulties with smoke in their attempts to prepare meals as a result of wet firewood. This makes them demand improved cookstoves" (Male. Focus group one: Site 3, 2015).
>
> "In the rainy season, I use my kerosene stove because firewood is usually wet so it produces a lot of smoke. I don't use my three-stone fire during this time" (Female. Focus group one: Site 1, 2015).

As a result of these health-related issues, many households reported making a shift from their three-stone fireplaces to alternative fuel/stove combinations in the wet season. These examples highlight the drawbacks of tracking access to higher tier cooking systems use when information is only collected on the primary stove and fuel used as in these study villages, the answers may differ between the rainy and dry seasons.

4.5. Incorporating end-user preferences into stove interventions and SDG7 monitoring frameworks

Despite emphasis by the SDGs on promoting sustainability and inclusiveness, efforts to formulate global targets and tracking frameworks inevitably run the risk of making compromises in terms of their sensitivity to local context (Sesan 2014, Satterthwaite 2015). At the same time, top-down initiatives emphasizing the cost and time benefits of fuel efficient cookstoves or the health benefits of "clean" fuels and stoves have had limited success in promoting their widespread uptake and sustained use by users of traditional biomass stoves (Sesan 2014, Thurber et al. 2014). Even though access to finance can be an important enabling factor within the ICS sector (GIZ 2013), contemporary market-driven approaches have often failed to meet the needs and priorities of lower income groups that have low demand for ICS (Kshirsagar and Kalamkar 2014). As a result, the cost of purchasing an ICS - especially one that requires regular fuel purchases - may prove too high for many potential users.

As data from Benue state illustrates, this is especially true when households face a range of competing financial priorities (food, education, healthcare, transport) and may still obtain some of their biomass fuel free of cost. Also, as the Benue respondents and Nottingham bake-off cooks illustrate, end-user cooking preferences and priorities are spatially and culturally specific and often differ widely from more technology-oriented systems for classifying improved cooking systems (Troncoso et al. 2007, Sesan 2014, Ray et al. 2014). Indeed, some respondents reported abandoning technologically improved systems in favour of lower rung/tier systems (or poorly maintained higher rung/tier systems) that better met their needs. Echoing research by Masera et al. (2000), many respondents from Benue and the bake/cook-off events practiced stove and fuel "stacking" in response to fuel and stove access/costs, the technical characteristics of different cookstoves (e.g. the type of food being cooked, the size of pot they could accommodate, their cooking speed and the amount of smoke created) and broader cultural preferences for particular stove types.

Such trends indicate the need for cookstove initiatives to make greater use of more participatory approaches that seek to understand end-user priorities for different technologies and the factors that help to create demand for these in different socio-economic, socio-cultural and geographical contexts. In particular, greater emphasis on "software" (as opposed to hardware or technology-oriented) approaches and social marketing initiatives of the type successfully used in the sanitation sector has potential to better understand and target the priorities of different user groups. Indeed, discussions about how to adapt such approaches that took place at a workshop organized by GACC, USAID, Universidad Peruana Cayetano Heredia and the Swiss Tropical and Public Health Institute in 2015 (GACC et al. 2015a) indicate increasing interaction between the water, sanitation and hygiene (WASH) and energy sectors.

In part, this reflects acknowledgement that increasing demand for cleaner fuels and improved stoves among low-income biomass-dependent groups is likely to be particularly challenging; especially where biomass fuel can be gathered free of cost and national energy policies overlook its importance as a household energy source (Pachauri and Jiang 2008, Barnes et al. 2011, Ray et al. 2017). But despite the recognition that - like poor sanitation - HAP-related problems are unlikely to be ameliorated without a community-wide response, the scope to transfer the types of community-led participatory initiatives used successfully in the sanitation sector are seen as limited (GACC et al. 2015b). This is because it is assumed that participatory approaches used in CLTS will be difficult to replicate in the energy sector given that cooking with "dirty" fuel lacks the sense of disgust widely associated with poor sanitation. Another hindrance to non-solid fuel adoption is that knowledge of HAP-related health problems is often low among low-income groups with low levels of education. In Nigeria, this reflects the lack of health education programmes focused on HAP compared to those promoting improved sanitation or malaria prevention (Akintan 2014) coupled with low levels of female integration (as change agents or ICS entrepreneurs) within the ICS value chain (Sesan et al. 2019). Other barriers include the wider benefits associated with smoke amongst communities that - in the absence of other methods of food preservation - rely on it for curing fish or meat (Akintan 2014).

Nevertheless, as the bake/cook-off events and Benue fieldwork have shown, participatory approaches can be designed to promote knowledge-sharing about both end-user priorities and the technical characteristics of ICS. They can also be used to encourage analyses of the wider impacts of biomass fuel use with efforts made to encourage discussions of the time, missed work/education opportunities and danger (e.g. risk of attack from animals or humans) associated with gathering this and the groups most affected by this. Resource mapping or matrix ranking exercises are useful for highlighting wider environmental or cultural factors underlying availability of and preferences for different cooking fuels or stoves whilst providing insights on locally specific barriers or enablers for the adoption of different fuels or technologies. Likewise, wealth ranking exercises can be useful for identifying key indicators of social status that help to contextualize energy choices in relation to other household priorities, aspirations and cultural norms regarding cooking system use. Approaches focused more directly on enhancing community-level understandings of HAP-related health issues, meanwhile could help communities to make links with commonly experienced symptoms. Drawing more directly on the triggering elements of CLTS, it may even be possible to mobilize dissatisfaction and drudgery associated with cleaning sooty cooking pots and living spaces to enhance awareness of HAP-related respiratory problems and their community-level impacts.

With regard to the future monitoring of which areas and groups make primary use of clean fuels and cooking technologies, the ambitious targets associated with SDG7 clearly require multi-faceted tracking frameworks that allow disaggregated data to be collected. GTF data collection activities will therefore have to go well beyond the scope of existing DHS questionnaires which only ask about the type of fuel households mainly use for cooking and whether cooking takes place in the house, outdoors or in a separate building. To provide useful insights on the extent to which economic factors hinder a shift to non-solid fuel, data need to be collected on the cost, availability and quantity of different fuels used. As case studies from Benue and elsewhere illustrate (Ruiz-Mercato et al. 2011, Ruiz-Mercato and Masera 2015), data on primary fuel and cookstove use provide only a partial picture. Widespread stove and fuel stacking along with seasonal or price-related shifts in stove/fuel use (both up and down tiers/rungs) can have important implications for exposure to health issues connected to HAP; the assessment of which need more detailed information on stove use as well as the health benefits associated with improved biomass stoves, noting that a recent randomized controlled trial (Mortimer et al. 2016) found no evidence that an intervention comprising cleaner burning biomass-fuelled cookstoves reduced the risk of pneumonia in young children in rural Malawi.

In recognition of this and despite SDG 7.1.2's emphasis on primary fuels and cooking technologies, recent workshops designed to feed in to the development of monitoring approaches for SDG7 highlight the need to capture information on multiple stove use and variations in stove or fuel use by season (Ruiz-Mercado 2015). Although the GTF does seek to capture information on "convenience" attributes associated with acquiring fuel and time taken to prepare stoves for cooking, priorities identified by our bake/cook-off cooks and Benue respondents for controllable, adaptable and quick-cooking stoves that can cater for large family sizes are not monitored, despite their likely influence on fuel and cookstove choices (Concern Universal 2016, Loo et al. 2016). Attributes linked to stove affordability, meanwhile, are only monitored for stove tiers 4 and 5, despite their importance in hindering a shift from three stone fireplaces. Likewise, safety-related and indoor air quality attributes depend on the availability of ISO data emissions data which is mostly restricted to commercially-available stoves that are beyond the price range of many low-income biomass users. As a result, questions need to be asked about who benefits from testing and certification (Mukulu 2014, Karve 2014) as it is likely to increase the cost to end-users whilst decreasing the margins of small-scale producers that may be better able to adapt their stoves to end-user requirements.

In pursuing their respective efforts to promote the adoption of clean cookstoves and fuels in 100 million households and "a data revolution for the energy sector" (IEA and World Bank 2015: 30), GACC and the GTF therefore need to focus attention on understanding and seeking to address key barriers faced by the biomass dependent poor. At the same time, national governments need to pay greater attention to the importance of biomass for household energy needs whilst creating enabling environments for ICS and clean fuel uptake. The slow uptake of non-solid cooking fuel to date suggests that SDG7's goal to "ensure access to affordable, reliable, sustainable and modern energy for all" by 2030 urgently needs more effective user-focused approaches that seek to understand the spatially specific and culturally-rooted nature of cooking practices whilst seeking to facilitate locally acceptable and appropriate forms of behavioural change. Solutions are starting to be sought from successful approaches within the WASH sector (Graham, 2015; Rosenbaum, 2015) but adaptations of CLTS-style community-led participatory approaches that play on distaste for the dirt associated with non-"clean" biomass fuels could help to promote change at the scale needed for a significant shift to the adoption of modern cooking solution to occur.

4.6. References

Agarwal, B. (1986). Cold Hearths and Barren Slopes: Wood Fuel Crisis in the Third World. London: Zed Books.

Akintan, O.B. (2014). Socio-cultural Perceptions of Indoor Air Pollution among Rural Migrant Households in Ado Ekiti, Nigeria. Unpublished PhD dissertation. School of Geography. University of Nottingham.

Ali, O. J. and Victor, M. A. (2013). Assessment of Socio-economic Factors Affecting Household Charcoal use in Makurdi Urban Area of Benue State, Nigeria. Journal of Environmental Research and Management, 3(7), 0180–0188.

Arickal, B. and Khanna, A. (2015). 'Principles and challenges in scaling up CLTS: experiences from Madhya Pradesh, India'. Paper presented at the 38th WEDC International Conference, Loughborough University. July 2015.

Barnes, D., Openshaw, K., Smith, K. and van der Plas, R. (1994). What makes people cook with improved biomass stoves? A comparative international review of stove programmes. Energy Series Technical Paper 242. Washington, DC: The World Bank.

Barnes, D., Khandker, S.R. and Samad, H.A. (2011). Energy poverty in rural Bangladesh. Energy Policy, 39, 894–904.

Beyene, A.D. and Koch, S.F. (2013). Clean fuel-saving technology adoption in urban Ethiopia. Energy Economics, 36, 605–613.

Bielecki, C. and Wingenbach, G. (2014). Rethinking improved cookstove diffusion programs: a case study of social perceptions and cooking choices in rural Guatemala. Energy Policy, 66, 350–358.

Bruce, N., Perez-Padilla, R. and Albalak, R. (2000). Indoor air pollution in developing countries: a major environmental and public health challenge. Bulletin of the World Health Organization, 78, 1078–1092.

Bruce, N., Pope, D., Rehfuess, E., Balakrishnan, K., Adair-Rohani, H. and Dora, C. (2015). WHO indoor air quality guidelines on household fuel combustion: Strategy implications of new evidence on interventions and exposure–risk functions. Atmospheric Environment, 106, 451–457.

Concern Universal. (2016). 'Linking Energy with Social Welfare Programmes: Integration of the Chitetezo Mbaula into the Social Cash Transfer Programme'. Conference presentation given at the Malawi Cleaner Cooking Camp, March 2016.

Dapo, B. and Emmanuel, O. (2013). Charcoal versus other domestic cooking fuels: survey of factors influencing consumption in selected households of Benue State, Nigeria. Journal of Sustainable Development in Africa, 15, 25–37.

Eckholm, E. (1975). The other energy crisis: Firewood. Washington, DC: Worldwatch Institute.

Evans, B. (2005). Securing sanitation: The compelling case to address the crisis. Stockholm: Stockholm International Water Institute with World Health Organisation and Norwegian Agency for Development Cooperation.

GACC. (nd). Clean Cookstoves and Fuels are Critical to the Success of the Post-2015 Sustainable Development Agenda. http://cleancookstoves.org/binary-data/ATTACHMENT/file/000/000/192-1.pdf

GACC (2017a). 'About'. http://cleancookstoves.org/about/.

GACC (2017b). IWA Tiers of Performance. http://cleancookstoves.org/technology-and-fuels/standards/iwa-tiers-of-performance.html

GACC, USAID, UPCH and Swiss TPH. (2015). 'Beyond distribution: Ensuring and evaluating the adoption of clean cooking and its benefits'. Presentation from a workshop held in Lima, Peru, May 2015.

GIZ. (2013). GIZ HERA Coking Energy Compendium: A practical guidebook for implemnters of cooking energy innovation. Bonn: Deutsche Gesellschaft für Internationale Zusammenarbeit (GIZ) GmbH Poverty-oriented basic energy services (HERA).

Graham, J.P. (2015) 'Behavior Change Frameworks, Models and Techniques'. Paper presented at the workshop 'Beyond Distribution: Ensuring and Evaluating the Adoption of Clean Cooking and Its Benefits' Lima, Peru, May 2015.

Hanbar, R. and Karve, P. (2002). National programme on improved chulha of the Government of India: An overview. Energy for Sustainable Development, 6, 49–56

IEEE. (2014). Global Humanitarian Technology Conference. http://ieeexplore.ieee.org/stamp/stamp.jsp?arnumber=6970352

International Energy Agency (IEA) and World Bank. (2015). Sustainable energy for all 2015 - Progress toward sustainable energy. Washington, DC: The World Bank.

Jenkins, M. and Sugden, S. (2006). Rethinking sanitation: Lessons and innovation for sustainability and success in the new millennium. Human Development Report Office. Occasional Paper for the Human Development Report 2006. London: UNDP and LSHTM.

Jenkins, M.W. and Scott, B. (2007). Behavioral indicators of household decision-making and demand for sanitation and potential gains from social marketing in Ghana. Social Science and Medicine, 64(12), 2427–2442.

Jewitt, S. and Rahman, S. (2017). Energy poverty, institutional reform and challenges of sustainable development: the case of India. Progress in Development Studies, 17(2), 173–185.

Kar, K. (2003). Subsidy or self-respect? Participatory total community sanitation in Bangladesh. IDS Working Paper 184. Sussex: IDS.

Kar, K. and Pasteur, K. (2005). Subsidy or self-respect? Community led total sanitation. An update on recent developments. IDS Working Paper 257. Sussex: IDS.

Karve P. (2014). Helpline-expert response by Myra Mukulu and Priyadarshini Karve. Boiling Point, 64, 25.

Kshirsagar, M.P. and Kalamkar, V.R. (2014). A comprehensive review on biomass cookstoves and a systematic approach for modern cookstove design. Renewable and Sustainable Energy Reviews, 30, 580–603.

Kohlin, G., Sills, E.O., Pattanayak, S.K., and Wilfong, C. (2011). Energy, gender and development: What are the linkages? Where is the evidence? Washington, DC: The World Bank.

Loo, J.D., Hyseni, L., Ouda, R., Kroske, S., Nyagol, R., Sadumah, I., Bashin, M., Sage, M., Bruce, N., Pilishvili, T. and Stanistreet, D. (2016). User perspectives of characteristics of improved cookstoves from a field evaluation in Western Kenya. International Journal of Environmental Research and Public Health, 13, 167–180.

Lozier, M.J., Sircar, K., Christensen, B., Pillarisetti, A., Pennise, D., Bruce, N., Stanistreet, D., Naeher, L., Pilishvili, T., Loo Farrar, J., Sage, M., Nyagol,

R., Muoki, J., Wofchuck, T., and Yip, F. (2016). Use of temperature sensors to determine exclusivity of improved stove use and associated household air pollution reductions in Kenya. Environmental Science and Technology, 50(8), 4564–4571.

Lewis, J.J. and Pattanayak, S.K. (2012). Who adopts improved fuels and cookstoves? A systematic review. Environmental Health Perspectives, 120(5), 637–645.

Lim et al. (2012). A comparative risk assessment of burden of disease and injury attributable to 67 risk factors and risk factor clusters in 21 regions, 1990–2010: a systematic analysis for the Global Burden of Disease Study 2010. The Lancet, 380(9858), 2224–2260.

Masera, O.R., Saatkamp, B.D. and Kammen, D.M. (2000). From linear fuel switching to multiple cooking strategies: A critique and alternative to the energy ladder model. World Development, 28(12), 2083–2103.

Mortimer, K., Ndamala, C.B., Naunje, A.W., Malava, J., Katundu, C., Weston, W., Havens, D., Daniel Pope, D., Bruce, N.G., Nyirenda, M., Wang, D., Crampin, A., Grigg, J., Balmes, J. and Gordon, S.G. (2016). A cleaner burning biomass-fuelled cookstove intervention to prevent pneumonia in children under 5 years old in rural Malawi (the Cooking and Pneumonia Study): a cluster randomised controlled trial. The Lancet, 389(10065), 167–175.

Mosse, D. (2008). Authority, Gender and Knowledge: Theoretical Reflections on PRA. Development and Change, 25(3), 497–526.

Mukulu M. (2014). Helpline - Expert Response by Myra Mukulu and Priyadarshini Karve. Boiling Point, 64, 24.

NBS/CBN/NCC. (2011). Annual socio-economic survey on Nigeria. Abuja, Nigeria: Nigeria Bureau of Statistics-Central Bank of Nigeria-Nigerian Communication Commission collaborative study.

Nagothu U.S. (2001). Fuelwood and fodder extraction and deforestation: mainstream views in India discussed on the basis of data from the semi-arid region of Rajasthan. Geoforum, 32, 319–332.

O'Reilly, K. and Louis, E. (2014). The toilet tripod: Understanding successful sanitation in rural India. Health and Place, 29, 43–51.

Pachauri, S. and Jiang, L. (2008). The Household energy transition in India and China. Energy Policy, 36, 4022–4035.

Peal, A.J., Evans, B.E., and van der Voorden, C. (2010). Hygiene and sanitation software: An overview of approaches. Geneva: Water Supply and Sanitation Collaborative Council.

Ray, C., Clifford, M. and Jewitt, S. (2014). The introduction and uptake of improved cookstoves: Making sense of engineers, social scientists, barriers, markets and participation. Boiling Point, 64, 2–5.

Ray, C., Sesan, T., Clifford, M. and Jewitt, S. (2017). From barriers to enablers: Where next for improved cookstoves? Boiling Point, 69, 2–5.

Rehfuess, E.A., Puzzolo, E., Stanistreet, D., Pope, D. and Bruce, N.G. (2014). Enablers and barriers to large-scale uptake of improved solid fuel stoves: A systematic review. Environmental Health Perspectives, 122, 120–130.

Rosenbaum, J. (2015). 'Behavior change approaches to facilitate clean cooking and reduce HAP'. Paper presented at the workshop 'Beyond Distribution: Ensuring and Evaluating the Adoption of Clean Cooking and Its Benefits', Lima, Peru, May 2015.

Rosenthal, E. (2009). By degrees: Third-World stove soot is target in climate fight. The New York Times, 15.04.2009.

Ruiz-Mercato, I. (2015). 'Critical implications of fuel-device stacking for initial diagnosis, monitoring and evaluation of stove programs'. Paper presented at the workshop 'Beyond Distribution: Ensuring and Evaluating the Adoption of Clean Cooking and Its Benefits', Lima, Peru, May 2015.

Ruiz-Mercado, I., Masera, O., Zamora, H. and Smith, K.R. (2011). Adoption and sustained use of improved cookstoves. Energy Policy, 39, 7557–7566.

Ruiz-Mercado, I. and Masera, O. (2015). Patterns of stove use in the context of fuel–device stacking: rationale and implications. Ecohealth, 12(1), 42–56.

Satterthwaite, D. (2015). Editorial: Is it possible to reach low-income urban dwellers with good-quality sanitation? Environment and Urbanization, 27(1), 3–18.

Sesan, T. (2014). Global imperatives, local contingencies: An analysis of divergent priorities and dominant perspectives in stove development from the 1970s to date. Progress in Development Studies, 14(1), 3–20.

Sesan, T., Clifford, M., Jewitt, S. and Ray, C. (2019). "We Learnt that Being Together Would Give us a Voice": Gender Perspectives on the East African Improved-Cookstove Value Chain. Feminist Economics, 25(4), 240–266.

Simon, G.L. (2010). Mobilizing cookstoves for development: A dual adoption framework analysis of collaborative technology innovations in western India. Environment and Planning A, 42, 2011–2030.

Smith, K. (1989). Dialectics of improved stoves. Economic and Political Weekly, 11.03.1989.

Thurber, M.C., Phadke, H., Nagavarapu, S., Shrimali, G. and Zerriffi, H. (2014). 'Oorja' in India: Assessing a large-scale commercial distribution of advanced biomass stoves to households. Energy for Sustainable Development, 19, 138–150.

Treiber, M. U., Grimsby, L. K. and Aune, J. B. (2015). Reducing energy poverty through increasing choice of fuels and stoves in Kenya: Complementing the multiple fuel model. Energy for Sustainable Development, 27, 54–62.

Troncoso, K., Castillo, A., Masera, O. and Merino, L. (2007). Social perceptions about a technological innovation for fuelwood cooking: Case study in rural Mexico. Energy Policy, 35, 2799–2810.

UN (2016a) http://www.un.org/sustainabledevelopment/energy/

UN (2016b) http://unstats.un.org/sdgs/indicators/database/

van der Kroon, B. Brouwer, R. and van Beukering, P.J. (2013). The energy ladder: Theoretical myth or empirical truth? Results from a meta-analysis. Renewable and Sustainable Energy Review, 20, 504–513.

Venkataraman, C., Sagar, A.D., Habib, G., Lam, N. and Smith, K.R. (2010). The Indian national initiative for advanced Biomass cookstoves: The benefits of clean combustion. Energy for Sustainable Development, 14, 63–72.

5. Fishing for food and food for fish
Negotiating long-term, sustainable food and water resources in a transdisciplinary research project in Burkina Faso

Gabriele Slezak, Jan Sendzimir, Raymond Ouedraogo, Paul Meulenbroek, Moumini Savadogo, Colette Kabore, Adama Oueda, Patrice Toe, Henri Zerbo and Andreas Melcher

5.1 Research context

In response to threats of chronic water scarcity and episodes of severe and long-term drought, the government of Burkina Faso started to create a widely dispersed network of water storage facilities throughout the whole country in 1950. As fisheries, these reservoirs have also become important new sources of food (Petit et al. 2009, Venot et al. 2011). However, there exist several threats to the services, e.g. fish and water quality, that these artificial aquatic sources provide (CNID-B. 2010a&b, Mahé et al. 2005). The predominant ones are overfishing, intensive agricultural activities around the water resources and the process of sedimentation. Within the framework of the Burkinabè national development programme, attaining food security and providing drinking water are central to the government's national development policies and strategies. Thus, to establish a sustainable management of natural and human-made aquatic systems, the responsible Ministry of Agriculture, Water and Fish Resources in Burkina Faso started to develop an integrative water and fish management strategy, which required methods and tools for the standardized assessment of the water quality and ecological status of rivers (MAHRH 2003, 2006).

The implementation and further development of this strategy requires personnel trained in management and the science that underpins its tools of data collection and analysis. In the framework of Austrian development cooperation, a senior manager of capture fisheries and aquaculture at the former Ministry's General Directorate for Fish Resources (GDFR) was supervised at the University of Natural Resources and Life Sciences (BOKU) in Vienna for his Doctoral research on fisheries and water management in Burkina Faso. The main aim of this interdisciplinary work was a general assessment of fish stocks, catchments, anthropogenic impacts on water, local knowledge and ecological awareness in fishing communities. At the policy level the purpose was to outline key areas for future management policy in the fisheries sector. The field research emphasized the importance of local fishing communities' knowledge on fish and prospective fisheries in arid inland waters and showed a serious lack of data on biodiversity and on river health (Ouedraogo 2010).

5.1.1. The establishment of a transdisciplinary research project

Further support for implementation of the water and fish management strategy came from an international project on monitoring and managing sustainable fisheries and water bodies in Burkina Faso[1], SUSFISH - Sustainable Management of Water and Fish Resources in Burkina Faso. This project recognized the history of failure of development projects based only on technical and/ or scientific advances. In Burkina Faso examples of abandoned equipment and infrastructure (fishponds, refrigerators, fish-weighing scales, fish shops) to support the modernization of fisheries testify to this. Aside from a few examples of successful organization of local management capacity, for the most part, there are significant gaps between national and lower levels of governance in Burkina Faso. Briefly, a governance system that effectively functions from the central, national level out to the regional and local levels has yet to be established (Melcher et al. 2018). Often the link between laws and actual practices in the monitoring of fisheries is not based on lived practical experience. One major challenge in fact seems to be adapting legislation to everyday practice. But traditional institutions play a vital role in reaffirming the identity of communities reliant on aquatic ecosystems and thereby broadly influence water and fish management. However, as our research indicates, the current governance structure does not link and harmonize these co-created rules with national laws (Sanon et al. 2015).

---

1   SUSFISH - Sustainable Management of Water and Fish Resources in Burkina Faso (2011-2014) http://susfish.boku.ac.at/

This governance gap between law and practice could prove to be a key barrier to realizing the potential for fisheries to become sustainable in Burkina Faso. The principle idea was, to establish a solid basis of useful knowledge in the social as well as the natural sciences in order to assess the extent and impact of this gap. Initiatives to establish this knowledge base were jointly founded by natural scientists in academia as well as by government officials so as to explore the possibility of analyzing and then managing fisheries based on biophysical scientific evidence. A transdisciplinary and participative approach was selected in order to integrate multiple perspectives of academic, policy and local practice. But nevertheless, the whole project was designed within the framework of development cooperation and therefore the research agenda aimed to contribute scientific knowledge to the social, economic and political barriers and bridges to sustainable fisheries. The overall objective was to strengthen the institutional capacities of the Burkinabè partners in higher education, research and management. A network of natural and social scientists as well as stakeholders worked together for three years in joint fieldwork activities, workshops and policy formulation for sustainable management and monitoring strategies suitable to the environmental and political context of Burkina Faso.

5.1.2. Integrating practices of participatory research

Interdisciplinary work can reveal important cross-sectoral activities, interrelated power relations and hindering factors that play key roles in the value chain of the resource fish in Burkina Faso. However, it was necessary to understand more about the complex interconnectivities and relations of sociopolitical activities in the natural resources management sector. In addition, there was a need to integrate applied participatory research methods. Taking into account the lessons learned from the SUSFISH Consortium 2015, the project aimed to involve local actors as well as actors on the policy level in the research process. Together, they would work on questions of water management and on assessment methods based on fish in order to contribute to the analysis of processes. The inputs of local politicians and of decision makers both in the fishing communities and on a national scale were integrated in the data collection. Here, relevant data on fish, the environment and on pressures were gathered. Also an analysis of the relationship between different kinds of anthropogenic pressures, including overfishing, intensive agricultural activities around the water resources, pollution by fertilizers, manures and pesticides, and the dynamics in fish assemblages and in water quality were integrated in the data collection process. The concept of participatory and interdisciplinary research was manifold and focused on the following tools for cooperation:

- Joint data collection in several field trips with five traditional fishermen, doing applied research, participating in the sampling of fish, macro-invertebrates and environmental parameters
- Joint laboratory work, i.e. identification of fish and macro-invertebrates, and data analyses
- Group discussions and interviews with fishermen and women processors (fishmongers) during their regular assemblies of more than 600 sampling sites (75 waterbodies)
- Interviews with local fishermen on their ecological knowledge
- Individual and group interviews with representatives of a) the central government (general administration) and b) the local communities (locally elected people), the technical staff of rural development, the fisheries and water stakeholders in six of the thirteen administrative regions of the country
- Workshops on gender issues with decision makers in fisheries and women's organisations on a local level in all sampling areas
- Workshops and group discussions with researchers and policy makers for systematization/synthesis of results
- Two workshops with the research team for the integration of a gender sensitive approach to the research agenda. In smaller workshops the research focus was further expanded to look for interactions both within and between biophysical and non-biophysical disciplines.
- Public conferences with practitioners, decision makers and scientists. During the joint research process the overall aim was to establish a transdisciplinary knowledge basis with inputs from the natural and social sciences as well as from a diversity of non-academic experts both within and beyond Burkina Faso, such as fishmongers, fishermen, NGO representatives, business people. Open fora such as public conferences provided opportunities to evaluate and disseminate project results.

Based on research in several disciplines in the biophysical, social, economic and political areas, including the expertise of decision makers, practitioners and scientists, this project was designed as a transdisciplinary project with several partner organizations and a large number of individuals. It was implemented by a consortium of eight organisations with expertise in the areas of research, education and development (Figure 1).

Figure 1: Sustainable Management of Water and Fish Resources in Burkina Faso (SUSFISH) project design and involved partners

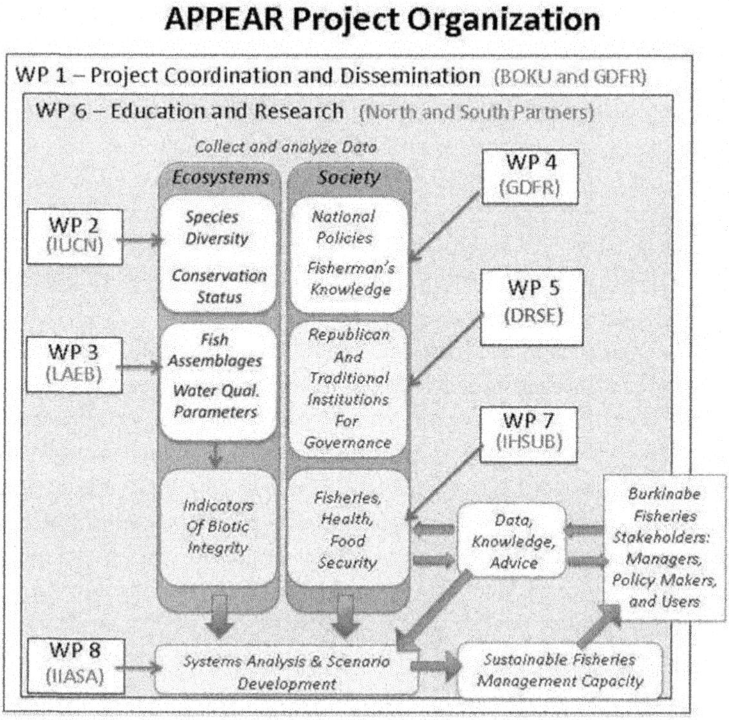

The design did not consider a direct interlink between the ecosystems and society. Therefore, the method of systems analysis was introduced to the core team as a tool for integrating data resulting from more than 20 studies in various disciplines.

5.1.3. Project results

The project's ecological and biophysical research results highlighted an important diversity of fish and macro invertebrates all over the country. More than 75 fish species and 105 taxa of macro invertebrates were identified and their spatial distribution and habitat characteristics were described. This diverse fauna occupies a wide range of habitat types. Habitats are always subject to human impact, but physico-chemical parameters are in accordance with tropical areas standards. Findings gave deeper insight into reactions of aquatic species to human pressures, specifically ranking species according to their tolerance to such pressures. The presence or absence of intolerant species can be key indicators of aquatic ecosystems under pressure and thus support ecosystem management at the landscape level.

At local level, fishing practices in Burkina Faso are very heterogeneous and depend on the status of water resources. Fishing communities relay mainly on rain-fed agriculture, irrigated gardening and cattle breeding as economic activities, and fishing as an additional source of income, so fishing is mostly a part-time job for farmers and herdsmen. Case studies revealed significant historical changes in colonial times when access to fish changed allowing foreign professional fishermen to exploit fisheries at a larger scale. But there still exist institutions of local fisheries management such as guidance by spiritual leaders, rituals such as collective fishing or closing times, as case studies revealed (Ouedraogo 2010, Sanon 2015). National fisheries government varies largely according to the status of water bodies. By law there are two kinds of status of fisheries based on management type. The PHIE are "nationally important" reservoirs, thus management is organized at a professional level and most of the fishermen involved are professionals, whereas concessions are more "subsistence-level" fishing for local markets. But the large majority of reservoirs in Burkina Faso are not included in these categories, do not have a legal status and the state is barely present to monitor and sanction illegal practices. In our field data, learning fishing methods over generations contextualized in the environmental challenges was found only in a few cases. Fishing techniques include cast nets, gill nets, long lines and traps. Even detrimental and prohibited methods like small mesh size and beach seines are very commonly used.

On the governmental level, findings indicate how development projects failed due to the limited ecological awareness of local stakeholders and a lack of training for local fishermen regarding fish stock issues. This raises the question of the availability of useful and reliable information, which is essential to proper decision-making for managing water and fish resources. Local communities as end-users need to have quick access to all available data on fisheries and water resources in Burkina Faso. However, little attention is paid to the way this information was communicated, what the local knowledge was like and therefore also the participation of different kinds of stakeholders in the process of policy formulation remained unclear. At local level the fisheries department does not cooperate with local communities in managing water and fish resources and on a formal level there is a lack of natural science fisheries experts in all of the 13 administrative regions. The legislation governing fisheries is well developed but is not translated into local languages and is thus not accessible for fishermen and fish processors. There is a serious need for a platform that involves all stakeholders at all levels (micro, meso and macro) to discuss such governance issues and meaningfully influence policy formulation and implementation. Studies from Zambia gave important insights how this process of developing regulations, that are crafted bottom-up, could provide by-laws that are addressing concerns and needs of local interest groups (Haller et al. 2016, Haller et al. 2018). This need is evident where official projects are monitored without the participation of direct stakeholders, and the main stakeholders working on the management level of their associations are not trained. A higher bargaining power could be stated at the mid-level of associations where fishermen are better organized and informed and have a higher income compared to self-employed fishermen.

The SUSFISH project's surveys in biophysics demonstrated that parameters such as fish size, abundance and diversity in Burkina Faso are related to the quality of fisheries and habitat management (Melcher et al. 2012, Stranzl 2014, Kaboré et al. 2016a, Kaboré 2016, Mano 2016). By using biological indicators, it became possible to distinguish impacted and non-impacted areas and to develop a multimeric index approach to assess the ecological quality of running water bodies in Burkina Faso. Thus, the project provided a rich data basis for monitoring the presence and impacts of pressures and developed technical tools such as software analysis and hardware for fish monitoring - including training in the use of those tools standardized scientific monitoring and assessment of the ecological quality status (e.g. electro fishing, Bentho Macro Invertebrates sample protocols, rapid field assessment protocol), applied to the environmental context in Burkina Faso.

Gender relevant research activities revealed that actors involved in fisheries are organized at several different levels of the units of production and by processing families. Men fish in groups while women are in charge of the processing. Another level of organization involves participation in provincial unions and the National Fishermen's Union of Burkina Faso. In this context, women, who - in most cases - do not fish on their own, control the area of processing. They play an important role in the exploitation of water and fish resources, because they allow not only the conservation of water resources, but also the survival and the community's reliability on fishing by developing strategies for small-scale distribution, transport and financing. For instance, women act as donors of loans for the fishermen in periods of financial distress.[2] But this predominance of women in the fish-processing domain does not translate into significant higher bargaining power regarding women's involvement in decision-making processes in administrative and legal areas of the fisheries management. Thus, the development of institutions for sustainable use of the fisheries should consider women as key stakeholders in the economics of fishery businesses.

5.1.4 Issues with the participatory approach

As previously mentioned, the long history of technically-focused natural science projects in fisheries that failed for social, economic and/or political reasons, prompted a transdisciplinary approach. This includes indigenous ecological knowledge and integrates biophysical as well as non-biophysical factors that might help (or hinder) the sustainability of fisheries in Burkina Faso (Sally et al. 2011, see also Haller and Merten 2008, 2018, Chabwela and Haller 2010 for fisheries in Zambia). Haller et al. (2016) emphasize that for a sustainable bottom-up institution building approach, local power asymmetries need to be understood as well as local knowledge needs to be incorporated (see also Berkes 1999 on this issue of scientific and local ecological knowledge and their differences and similarities). Thus, a participatory approach accesses multiple, non-academic perspectives that are vital to transdisciplinary research (Thompson Klein 2004, Haller et al. 2016).

In SUSFISH this became evident, since the nature of problems to be addressed were not per se in the field of natural science but rather in the social,

---

2   Our studies revealed that 82% of fishermen have already contracted loans with which to purchase fishing equipment and fishing licenses. In 88 % of cases, the loans granted by women are repaid in kind (fish) from the production.

political and institutional context and thus needed to be negotiated cooperatively by experts from both science and (political) practice. The project sustained participatory approaches by scientific experts and policy makers collaborating in workshops with local partners as local experts to develop joint strategies to communicate the results of the scientific cooperation to local communities. This involvement of local partners as experts to share and apply their knowledge was central towards a process of shared research. This, of course, is intimately linked to language, so as to contextualize the findings of the natural science research and make it more accessible and, as a consequence, more applicable in local fishing communities around Burkina Faso. In the tradition of internationally oriented, policy-relevant science, the project sustained three levels of knowledge sharing and learning activities: 1) joint fieldwork, 2) policy formulation and strategic sector development activities for fisheries and water, and 3) synthesis of research results and definition of lessons learned including important remaining questions. But important to note, within this framework there was no space reserved for the direct involvement of local groups to deal with the problem and to start a joint bottom-up process, such as concerns-oriented drafting of by-laws described for Zambia in Haller and Merten (2018) for instance. Additionally, local actors in the various fishing communities were not a homogeneous group. Besides the diversity of socio-historical contexts, their knowledge is linked not only to French, but to local languages such as Jula and Moore as well, which were very important linguistic resources for interaction and joint discussions.

5.1.5. Fieldwork - practice and training

In order to elicit and share knowledge, interactions during fieldwork activities between scientists, policy makers, local decision makers and practitioners in the fishing sector were continuously encouraged. They started in the beginning of the project and were continued during the whole time and have been mainly conducted by a large group of students participating in the research. Developing an innovative and likewise adaptive assessment of the integrity and long-term sustainability of water quality and fisheries in Burkina Faso required intensive fieldwork for sampling and data collection. Within the project's framework of capacity building in higher education[3], a large number

---

3 APPEAR is a programme of the Austrian Development Cooperation (ADC) to support higher education and research for development on an academic institutional level in

of Burkinabè and Austrian students carried out most of the data collection. During their participation in several field trips for joint collection of data, students learned to work in teams together with junior and senior researchers in different study areas all over the country. Subsequently they continued the work independently. These joint excursions and field trips were designed, organized and implemented particularly with the participation of practitioners.

For instance, the 21 study areas were selected by the students' supervising scientists in collaboration with government officials. Besides biophysical and ecological criteria it was important to take into account the local management practices influencing the condition of water bodies and fish stock. Therefore, students conducting their field research were assisted additionally by supervising staff, which included traditional local fishermen, local government officials of both the republican and traditional systems and representatives of women's associations. This assistance by local experts happened either while participating in the sampling of fish, macro-invertebrates and environmental parameters, or during ethnographic fieldwork such as participant observation, interviewing and group discussions. Local experts were selected on the basis of professional experience (e.g. fishing techniques, years of training), access to local and state institutions, ecological knowledge and personal availability for the series of field trips. The local fishermen more and more shaped the fieldwork practice. Although every student followed her or his individual research focus, the project staff organized joint workshops on fieldwork methodology (within their disciplines as well as interdisciplinary).

In terms of training in ecological and biodiversity approaches, scholars at various levels of their academic career (BA, MA and PhD) were trained in seminars and in the field to undertake surveys of fish and benthic macro-invertebrates. Nine study areas all over the country were visited together during four sampling campaigns. Joint activities involved students in developing and adopting standardized field collection techniques, species identification and enumeration methods, as well as in analyses using aggregated biological attributes or quantification of key species. For instance, in the beginning it was necessary to develop a joint fishing technique, which provided a standardized

---

the ADC's southern priority countries and key regions and in three priority countries of the South Caucasus and Black Sea Region. It provides funding for academic partnerships between higher education and research institutions in the addressed countries and Austria. https://appear.at/en/

method for sampling in the heterogeneous landscape of Burkina Faso's water bodies. The best way how to negotiate and ultimately, agree was a joint training of the whole group. The outcome was a threefold method, which was oriented towards local experience in fishing: every site (segment) was sampled by electrofishing, traditional fishing methods (cast and gill nets, active and passive fishing) and fishing for benthos. Even the frequency and extent of sampling in the individual areas had to be negotiated, as it was not the goal to fish as much as possible, but to get a random result. It was important to use the appropriate fishing method for the investigated water bodies to get meaningful and consistent results on diversity and abundances. But this approach was challenging also in terms of communication: after one month of working together the students started to realize that communication was crucial to this process and that knowledge not only culturally evolved but exists as knowledge-practice-beliefs complexes embedded in their institutional contexts. It was an important experience for them to learn about alternative knowledge and perspectives based on own locally developed practices of resource use (Berkes 1999, Haller et al. 2018). Translation should not stick to a word-by-word transfer, but rather to translate concepts such as for instance that of understanding biodiversity. This is well illustrated in the following example from a cast net (Figure 2) event described in the research diary section below by Paul Meulenbroek on 12.12.2012 in the catchment of Nazinga:

> "During our fieldwork, electrofishing and cast net fishing were approximately conducted the same time frame and both methods captured around 9,000 specimen, 18 species were only caught with electrofishing and 11 exclusively with cast net. For the latter experienced fishermen are needed to perform effective fishing. The most impressive demonstration of the professional fisherman Noufou Bonkoungou's fishing experience was performed in the protected area of Nazinga. First, he walked along the river for about 10 minutes, suddenly he stopped to wait without any motions for another 5 minutes and starred at the water. Unexpectedly he started to run and throw his cast net. He could catch more than 700 specimens with one throw. In comparison, it took the students 3 hours to get one single fish. It was a very important experience for students, how to evaluate local knowledge, which is often implicit and not considered by the scientific experts sufficiently as data for their analysis."

Furthermore, the whole group worked with standardized field protocols drawn from European and international research experience. Because of

Figure 2: Electrofishing and cast net fishing. Picture: Paul Meulenbroek

fishermen's input, the participants realized during the sampling phase that they needed to change the standardized tool by integrating new parameters such as abiotic factors (e.g. extensive farming practice) in order to adapt fish and invertebrate habitat assessment protocols to the conditions in Burkina Faso. It was important to take into account many country specific factors such as intermittent rivers, mainly man-made waterbodies or a country specific velocity of rivers. A specific challenge for the team was the development of a classification tool for temporary streams, as they dry out during the hot season. Due to the participation of local fishermen, the selection process of segments of reservoirs for sampling considered local ecological knowledge. This ongoing process of recognition of "non-scientific" knowledge systems lasted for several months and integrated the inputs of local experts such as the fishermen, local authorities of the traditional government and representatives of fishermen's associations, who all contributed their experience on seasonality of the water body and fish migration practices. It created a zone of learning and understanding on both sides, among the scientists and the local actors. Comparable to other areas a solid basis of trust was created, which invited on local level to participate, share and co-create in research activities (Haller et al. 2016: 82).

Here it is important to note that this kind of interaction was only possible because members of the scientific team could make local knowledge accessible through linguistic translation. Because of their linguistic background and their academic education, they were able to interpret this local knowledge into scientific parameters understandable to the whole team.

Local knowledge was better integrated with international science by performing different stages of data collection, processing and analysis jointly in the field. Fieldwork was embedded in joint activities to prepare specimens and process data after collection, such as laboratory work to identify and control quality for fish and benthic-invertebrates taxonomy. As local fishermen were also directly involved in these accompanying measures, the description and classification of fish species happened *in situ* at the sampling sites. Activities on sites were documented in field protocols and diaries in order to link results to workshops on the utilization of Red List criteria and categories[4]. Finally, the local knowledge on how to interpret features and characteristics of fish contributed significantly to the evaluation of the conservation status of fish species at national level. These workshops did not take place in the field but in a scientific environment framed by academic expertise. It would have been an important contribution to the evaluation process if we had considered local fishermen's expertise also at this level of research activities. Our internalized division between science and local ecological knowledge did not

---

4   An official list of fish species and invertebrates and a national database of metainformation on existing biophysical characteristics of fisheries, the diversity and conservation status of fish species and benthic invertebrates, the pressures on fish populations and methods of water assessment based on fish and macro invertebrates was developed during project time.

allow the imagination of the importance of such a contribution. But by reflecting this collaborative process at a later stage, we addressed the excluding effect of drawing on our existing constructions of the field, informant and science (see Foucault 1981).

Second, it clearly showed the limits of imagination of transdisciplinary fieldwork. The trips were planned along disciplinary boundaries. The focus was on ecological and biological data collection and therefore students trained in participant observation and interviewing techniques were not involved. All these important reciprocal processes of knowledge co-creation should have been elicited and processed with socio-anthropological methods of qualitative research for a better understanding of specific institutions governing the use of fisheries, such as the notion of spiritual ownership of water bodies or rules and regulations originating from pre-colonial times (see also Haller and Merten 2010).

In terms of training in socio-political approaches, students worked on issues of fisheries management, governance, society and local fishing as well as fish processing practices within the larger framework of the institutional cooperation of several university departments and the national government. Joint training workshops on research practice exposed them in two stages to transdisciplinary and explorative approaches in social, economic and political science: first, by developing a set of quantitative and qualitative data collection methods, and second, by subsequently coordinating and adjusting data analysis to the team's focal research interests.

Experts from various academic and political organizations (project team members, Work Package leaders or supervisors) provided flexible supervision in accordance with the particular implications and dynamics of the applied participatory approach. Gender expertise, for example, was integrated in the research agenda by workshops for junior and senior scientists. This included training for students on gender sensitive field collection techniques, such as methods for identifying factors for gender imbalances, implementation of participatory and awareness raising strategies in fieldwork as well as on focusing on gender issues in data analyses. The expert's long-term practical experience working with women's organizations provided methodological knowledge of how to include women into the research process although they are not represented by associations or in decision-making. In addition to these workshops, during their research she continuously reviewed their written materials e.g. questionnaires, results and reports, which she shared as part of group discussions in joint meetings. This enhanced the students' critical self-reflection of their own research practice and of the scientific paradigm they were working in. They became more sensitive for power relations and questions of equality in the research process and reflected their own role as researchers. They had to think about their expectations in the collaboration with local actors. As a result, the field studies provided by these students feature an outstanding interest on gender issues in fisheries, nutrition, health and natural resource management, which contributed generally to applied gender research in Burkina Faso.

To conclude with an example, it was an important contribution of the government's gender expert to share her knowledge on key issues on socio-economic aspects in the water and fisheries management with the students. As a team member she was able to bring in her experience and thus improve research findings. Her contribution included taking into account gender sensitive factors for the composition of focus groups or issues of female representation on family and community level; a critical revision of questionnaires and data in terms of economic responsibilities of women in fishing communities; considering the neglect of female actors in the fisheries on policy level.

Gender sensitive studies revealed that female actors play an important economic role in the exploitation of water resources, because they support not only the conservation of water resources, but also the community's survival and their reliance on fishing. In addition to carrying out their activities, women transformed themselves into donors of loans to the fishermen, who in fact monopolize the commercial fish trade. As a consequence, women are economically important stakeholders in fisheries, but structurally excluded from decision-making processes. Findings showed that women's involvement in fisheries helps improve both diet quality and, especially, households' food security during the year.

5.1.6. Involvement of policy makers - key questions of management

Second, participation happened at local[5] and regional levels during the process of policy formulation for fisheries and water management in the context of workshops and meetings with representatives of water authorities, fishermen's associations and local governments. Three experts of the GDFR (Minis-

---

5   At local level, officials involved in the implementation of national policies and strategies for fisheries management in Banfora, Tiéfora, Cascades and Moussodougou have been included.

try of Animal and Water Resources Management), who were part of the project team, guided this process, which should lead to a new strategic orientation of the Burkinabè fisheries sector.

In these workshops and meetings, it was their aim to link ongoing research findings to the level of government's interaction with local and regional officials, technicians and representatives, who were in charge of training local fishing communities. In a reciprocal process, the local and regional perspectives should help to improve government policy. The interactions were therefore not limited to data collection. In field surveys the constant contact with the various actors sometimes questioned the research methods and experiences and required adjustments. For example, at the beginning of the study we did not include the local agricultural officers. This group has been suggested by the manager of the fisheries and was later integrated. Fishermen and fishmongers mainly organized the meetings of focus groups. They mobilized themselves and determined the appropriate period of time to bring the group together.

To illustrate how this interaction happened, we refer again to gender issues in the fisheries and water management policy as an example. In 2007, the General Directorate for Fish Resources of the Ministry of Animal and Fish Resources in Burkina Faso assigned a water engineer and gender expert to assess female representation and gender imbalances in the fisheries and water sector. Her work continued in the framework of the SUSFISH project focusing on the set up of a process of policy formulation. At the end of the project, a final draft of a „Stratégie d'intégration du genre à la politique de développement et de gestion durable dela pêche au Burkina Faso" was adopted by the Ministry.

First of all, this participatory approach required the facilitation of communication within the multinational team of researchers and practitioners. Secondly, in order to reflect on the input of local experts regarding fish stock, natural diversity and fishing her team members had to develop a joint strategy not only to communicate the results of the scientific cooperation to local people, but also to involve them as experts in the setting. To lower the risk that her team acted in ways that are alien and incomprehensible to local people's life worlds, unmediated interactions with local actors were organized by local technicians in local languages. In the framework of government's campaigns and technical training workshops all over the country data collection were organized with practitioners such as technicians in the fisheries, aquaculture and in water management. Their knowledge should influence the findings of the natural sciences research. The ministerial expert monitored these interactions in separate trainings and evaluation meetings based on a gender sensitive approach. She helped them to take into account the main barriers for women to access local resources, written information or legal rights. A further topic was how to work interactively with women in male-dominated surroundings and that group composition for instance can influence if women will take the opportunity to speak for themselves. It was a major objective to achieve a deeper understanding of complex underlying processes and correlations in socioeconomic and "socio-ecologic" systems through this specific involvement of local experts of the fisheries and water management policy.

5.1.7. Synthesis of research results

During its final year the project tested ways on how to integrate scientific research in policy making and broaden the knowledge base by incorporating the perspectives of people acting in different sectors at different levels of society, from local to regional to national. This meant expanding the scope of research and policy discussion beyond interdisciplinarity (social scientists, biologists, fish ecologists and nutritionists) to include inputs from outside academia: managers (fisheries planners, policy makers) and practitioners (fishermen, fishmongers, traders) at the local and regional level. The main aim was a synthesis of all findings for final reporting and to develop key research questions for further projects. As a follow-up it was planned to formulate policy briefs for the government of Burkina Faso.

For this part of the research process, a member of the consortium provided expertise in innovative methods of systems analysis. The method was introduced to the core team as a tool for integrating data resulting from more than 20 studies in various disciplines. In two series of workshops in November 2013 and July 2014, the process assembled a diverse focus group consisting of academics and government members, who collaborated in the design and running of an experiment in scenario development (Sendzimir et al. 2011, Ouedraogo et al. 2014). Tools such as scenario development and system analysis were applied in workshops and modeling sessions. Time constraints often deny policy makers and stakeholders extended opportunities to explore the future. Therefore, the challenge for all members was to do so in a very short time period. This was done by defining the variables and their relationships that may influence future development pathways for fisheries in Burkina Faso. This exercise developed scenarios as ways for experts and partners to examine the dynamic implications of the facts and questions generated by the project. Such exercises allowed participants for the first time to bring together, discuss and reconsider their assumptions and questions in light of the dynamics that they anticipated. It further allowed an elaboration of what particular variables and parameters ought to be measured in order to better understand how the socio-political and ecological system is changing. In order to understand the phenomenon of water and fish resource management in Burkina Faso holistically, we developed, changed, and jointly adapted our concept in an open-ended process. In terms of research practice, this was very challenging to organisation, communication, and integration of heterogeneous types of knowledge, as explained in the following.

Challenge of a complex process: The process of conceptual mapping exercises consists of several steps. It starts with the telling of a story as well as the formulation of hypotheses and key questions and continues with the ranking of the influencing parameters. In our workshops the joint examination of possible structures of relations that might underlie the dynamics opened up space for debates. These efforts were intentional and helped us to identify how some partners imagine concepts or patterns of relations and how they would propose to analyze them in the future. But it became evident that it is a challenge to examine phenomena through the broader lenses of inter- and transdisciplinarity. Frequently we found that it requires patience, trust and encouragement to embrace the complexity of problems. This means resisting the historical dependence on explaining problems by means of single, key variables and sustaining our mutual examination of the dimensions of that complexity, arriving at more nuanced understanding of multiple causation. We learned to deal with the complexity of differences but also to consider their scope of negotiation (Bhabha 2000). The process of ranking parameters, for instance, follows a linear structure that requires defining which parameter comes first and therefore has more analytical weight than another.

Challenge of disciplinary boundaries and epistemology: At the beginning of the project the differences that distinguish how each discipline focuses its research were evident to team members. During the first half of the project, these distinctions defined how we worked and were followed more or less separately. On a technical level, the project design included reporting and exchange of information for monitoring and evaluation but did not integrate a deeper understanding and learning process transdisciplinarily in all research activities. This latter reflexive process was limited to some few workshop and fieldwork contexts (Figure 3), as mentioned above, where room for dialo-

Figure 3: Workshop on systems analysis in Ouagadougou, June 2014

Thinking and reflecting in a group of up to 20 participants means to open up the floor to many, often diverging, interpretations of the specific parameters. Therefore, it was sometimes difficult for us to find a compromise for defining parameters. However, on the other hand this participatory method enhanced our ability to reflect on how we mutually collaborate to share and combine our understanding and follow correlations to define future research and policy.

gue and exchange was provided selectively. The final group discussions on our project results issuing from natural and social sciences views as well as from a diversity of non-academic sources were intensive and not always smooth because of disruptions, rejections, misunderstandings, and ideologically loaded conflicts. However, overall, we experienced ourselves as part of a transdisciplinary team with complementary bodies of knowledge and ways of knowing, - lay, local, and indigenous expertise full of contradictions, which do not lead to simplification.

Challenge of flow of knowledge and information: Our project's goal was to explore ideas that are risky because their implications lie so far into the future that prediction becomes impossible. To this end the process assembled a focus group comprising a diversity of academics and government members with the purpose to apply forecasting and back-casting techniques (Kok et al.

2011), wherein we developed various scenarios in different development exercises. These included extended discussions and controversial negotiations on definitions of sustainability, governance, subsistence and gender concepts between the involved experts (see Bourdieu 1996). The exercises provided a problem-oriented framework that invited creative ways of thinking. Consequently, it enabled a team of experts to enhance their own understanding of how they saw problems and contexts as well as how other stakeholders did. In this way the negotiation process brought even more complexity to light in a way useful to the network working on these problems. By sharing knowledge and understanding as a group we developed a collaborative knowledge base that had not been established yet.

For practical reasons this collaborative process did not involve all of the other actors engaged at earlier stages of the research. Even the group of students, who played a significant role in data collection and processing, were not included in the team. As outlined above, types of knowledge were elicited and negotiated at various steps of the research process. It would have been very important for the core team to ensure that it could become integrated in this last step of comprehensive synthesis, as we dealt with a very heterogeneous group of participants and various forms of knowledge. In fact, we, as members of the core team, became important bearers of knowledge, but were not fully aware of its importance and struggled to bring together the perspectives of diverse stakeholders, especially policy makers, local authorities and practitioners. Referring to the project's approach, the knowledge integrating process in its complexity was clearly underestimated. The various scenarios developed did identify many potential barriers to making fisheries sustainable, such as de facto open-access to fisheries, increasing pressure through agriculture and mining practices, disruption of fish migration by dams, dysfunctional government institutions because of lack of state financial support for monitoring and sanctioning, the policy focus on a few large reservoirs of „national economic interest" (Melcher et al. 2018: 530), and very heterogeneous local communities in terms of internal power distribution, few interaction in dealing with outside actors and little bargaining power as well as top-down developed law and practices to monitor fisheries.

However, it would have been a significant addition to the project to rigorously identify how to exploit opportunities and circumvent challenges over the next 20 years in order to successfully establish fisheries science and management in Burkina Faso. Not only scientific and political representatives, but also students and other local actors such as fishermen or fishmongers should have been part of the debate. These tasks clearly belong at the top of any future research agenda for Burkinabè fisheries and water management.

Challenge of time: There was insufficient time to sustain these exercises long enough to allow stakeholders to fully apply the knowledge bodies and to explore different pathways to the development of sustainable fisheries over the coming decades, such as applying forecasting and back-casting techniques to exploit opportunities and challenges for the next twenty years together. In these workshops, we needed at least three times as much time to establish a transdisciplinary knowledge base for further work on sustainable fisheries. Because of our extensive discussions, which lasted several days instead of the two provisioned for, the first series of workshops ended with the ranking of parameters. Altogether, the transdisciplinary project team would have needed more time and virtual space to continue the process of evaluation and understanding to the point where policy recommendations could have been identified and agreed upon by the relevant stakeholders. The output of this process did not include the activity for participatory scenario development. But our exercises of systems analysis provided an interim base of research findings, which derived from the identifying key parameters and key relationships between them. As we had to adapt the research agenda, the modelling session had been pushed back. So, the team as a whole could not run through the creative process of developing new scenarios and exploring the various pathways they embodied. Reasons for this decision were a) significant restrictions by (non)availability of team members and travel costs, b) the project schedule was very tight and important milestones, such as the delivery of models, were already late. Therefore, the modelling had to be introduced and organized in a one-day workshop and was left to a small team of Austrian team members, which completed the experiment of modelling (see Figure 4).

Challenge of translation and language: In addition to the time factor, a second major challenge was to address the plurality of languages and linguistic resources during scenario development exercises. In terms of languages employed during fieldwork, the team used plurilingual practices, such as translation and interpreting, in order to foster dialogue between researchers, practitioners and the local community on sustainable fisheries policy. As previously shown, using plurilingual practices was very important to allow a dialogue where all participants were on a par with each other. The process of translation was more than the transfer of knowledge from one word to another. Rather it required a critical approach to how we frame and codify knowledge within our disciplines and to learn how others do it. This became evident during the

Figure 4: SUSFISH research results related to food security, sustainable fisheries, water quality boost the potential for development and education in Burkina Faso

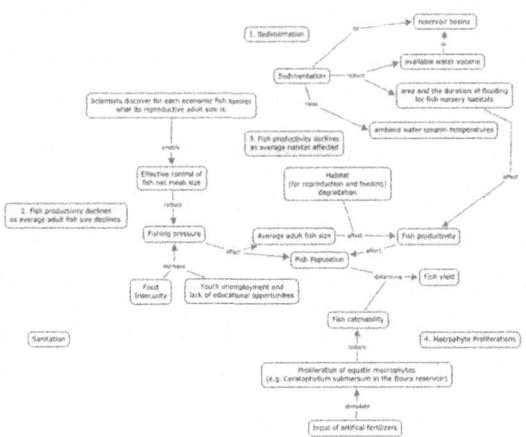

whole research process, but explicitly in the interdisciplinary adaptive management workshops.

The involvement of diverse linguistic resources implies a reflection of prejudices, ideologies, and of understanding roles that are associated with different linguistic cultural backgrounds. It also meant to reflect on the exertion of various different symbolic power and inherent power relations the team was not aware of. The predominant use of English as language of scientific communication risked to become a barrier to open discussions at some levels of interaction. The questionnaires, for instance, used scientific and technologically influenced language in the elaboration of key questions and in the applied parameters. In our team, scientists and practitioners adapted quite differently to this kind of knowledge translation applying a terminology with key parameters using numeric values for modelling. The project idea was embedded in the frame of bio-physical research, which provided the leading ideology in defining relevant factors. We added socio-political questions, but these did not constitute the source for developing ways to reflect. For our social scientists it was a learning process to search for the linguistic terms to tell these scientifically framed narratives. It is a question of epistemological positioning; knowledge was much more contextualized in the French, more hegemonically grounded scientific tradition and therefore in a distinct tradition of narration. Therefore, it was important to develop a sensitive strategy to avoid the temporary exclusion of team members in the transdisciplinary discussions. We experienced for instance that it helped that a small group switched to French or Moore or German to clarify certain points before continuing the discussion in the whole group.

Challenges on an organisational level: As the workshops were highly participatory, this approach needed the commitment and collaboration of every group member. But in practice, it was very difficult to reach this goal in a group of 20 persons, who are partly in key management positions and changing working-conditions. Addressing the challenge of spontaneous changes of availability, the responsible team members tried to focus the joint discussions on issues following the disciplinary divide of bio-physics and socio-political sciences. It became evident that those discussions, which were held as plenary and thus interdisciplinarity, were much richer in terms of comprehensive input and shared learning.

5.1.8. SUSFISH's participatory approach: lessons learned and problems

Overall, the project team achieved important progress in generating knowledge in terms of new concepts, facts and perspectives about aquatic ecology, water quality assessment methods as well as fisheries and water management practices in all 13 regions of Burkina Faso.

The two PhD students from Burkina Faso, who were also partly studying in Austria, and the four Austrian MA students, who studied in Burkina Faso, as well as the 13 Burkinabè MA students in Burkina Faso, formed a group of young scholars who wanted to reflect on different research contexts. They spent many weeks together, discussing, analyzing, and learning. Together they had the opportunity to experience practical constraints and to learn how to integrate this transcultural and interdisciplinary experience into the research process. At the end of SUSFISH, the first generation of publications and theses show an impressive contribution of junior scientists based on the joint and reflexive research process of collecting data during fieldwork. Anyhow, the time when all students from every discipline were together was too short for them to deeply learn and to benefit from other perspectives. Future projects should focus on providing greater opportunities for such interdisciplinary learning.

The 19 students, responsible for data collection and processing, interacted the most with practitioners during their fieldwork. Many of these relationships between students and practitioners were long lasting and included an intensive, highly participative joint research process. Our project design was flexible enough and would have allowed for their participation in the process of systems analysis. We could have incorporated them, when we designed events. But actually, their participation was prevented by several reasons: It was generally hard to assemble the whole team, we would have needed several more meetings just with the students to include the diverse perspectives from such a large group. The flexible design ran into constraints of time and money. Hierarchical aspects of supervisor-student relations did not made room for their active participation. In terms of project design, they were not explicitly integrated on paper. However, we could have insisted on this from the beginning. Their participation in this crucial part of the project would have enriched our discussions and improved the process of systematization of knowledge. Including students and other local actors, such as fishermen or fishmongers, in addition to the scientific and political representatives in the debate, would have been a significant contribution to rigorously identify how to exploit opportunities and circumvent challenges to a successful establishment of fisheries science and management in Burkina Faso over the next 20 years.

The participation of local authorities and practitioners from the fishing communities, in joint research activities was intensive in terms of knowledge elicitation and data collection on a local level, but not in the evaluation of research findings. Had the project lasted longer, a reviewing process of SUS-FISH findings with local actors would have been a significant contribution from both groups. One goal of the project was a "communal review" to share outcomes and information with actors on the local and regional levels. For reasons of time constraints and project monitoring we did not succeed in this point. With this communal review, we possibly could have achieved our goal. But it is also a matter of fact that for instance data processing in a scientific environment framed by academic expertise is hardly accessible for local actors. Therefore, alternative ways of interaction are needed. For example, it could be an interactive workshop for evaluation, where local fishermen's expertise can be considered even at this level of research activities. The final symposium in Ouagadougou, which was held as a public event, was a great experience in terms of examination by practitioners. As delegates from 36 associations and organizations at local, regional and national level were present and participated vividly in the debate, the challenge for us is how to continue such multi-level exchange after the project ends.

The interaction between students and policy makers in the frame of gender methodology workshops contributed significantly to the transdisciplinary approach of SUSFISH in comparison to the workshops, which were held within disciplinary boundaries. These joint meetings were on site and enabled mutual understanding and joint learning, as the gender related activities showed. Furthermore, the reviewing activities (fieldwork tool kits, questionnaires, results and reports) of the gender expert became a source for defining further questions. The research team became aware of this important contribution and implemented these evaluating measures for other project reports. But time was too short to cope with the organization of a joint evaluation workshop at academic level before the project end.

For a synthesis of research findings, scenario development and systems analysis were applied in workshops and modeling sessions in a very short time period during the project's last year. These tools require substantial inputs from the whole project team, because they entail complex processes of reflection and discussion to address interdisciplinary questions. Sustaining such discussions requires full commitment by participants and cannot be managed by a single person. The process of negotiating meaning and mutual understanding in a highly transdisciplinary context was enriching and improved the joint knowledge base, but it was a challenge for all members to link the various components in a very short time period. This reflexive discussion successfully led to the formulation of a strategy discussed in a presentation for ministerial stakeholders and decision makers. However, SUSFISH ended when dialogue with politicians started, a point when the project's collective experience should be applied while data is current and most useful. Future projects should initiate such dialogues in an earlier phase. The final phase should extend them to jointly develop a set of strategies with policy makers and practitioners as part of identifying scenarios to successfully establish sustainable fisheries management.

Overall, the synthesis was an important contribution of the project to initiate this process of learning and understanding. The outcome was two sets of lessons learned and remaining open questions that can help define future research agendas. The process of knowledge transformation and understanding is still ongoing, and was an important part of SUSFISH work. But to all of us, it was an experiment to use tools such as scenario development for it, and we learned, that these interactions need to be considered appropriately in the research design in terms of expenditure in time and costs.

Participatory research enables continuous learning. We experienced very different modes of learning at those various layers where participatory exchange of knowledge happened in our project. To illustrate how and by whom meaning was co-created and our understanding enhanced, we will refer in the next section to selected moments of our joint process of systematization of knowledge and reviewing our findings. The gender sensitive approach introduced to SUSFISH will serve as example, as it provoked contestation because of inherent power structures and a donor-driven ideology. Questions of language resources are crucial, when it comes to debate and contestation. In SUSFISH we used plurilingual practices, but as it revealed aspects of ideological positioning translation we will also point to translational practices and the role of communicative possibilities in the context of epistemological hegemonies.

5.2. Key moments of participatory research

5.2.1 Scenario development workshops - key to understanding

Predicting and managing a complex and evolving world is difficult. Adaptive management represents experiments in new ways to learn and adapt science, policy and practice even as one manages a socio-ecological system (Sendzimir et al. 2017). The imperative to flexibly manage in adaptive management prioritizes processes that generate learning, meaning, knowledge and experience of ecosystem dynamics (Folke et al. 2005). The SUSFISH project embedded such learning processes in a series of scenario development workshops. In these exercises, team members representing academia and policy co-created knowledge about factors key to alternative scenarios of how fisheries management might be developed in Burkina Faso. Our team objectives in these workshops were to learn-by-doing participatory science to contribute to establish it as another option in the support of policy formulation and application in Burkina Faso. Therefore, the exercises allowed participants to mutually develop skills to analyze and communicate complex ideas and formulate policy as well as to study the factors influencing the dynamics of water and fisheries management in Burkina Faso (Peterson et al. 2003). By and large, the methodology adopted to develop exploratory storylines during the first workshop series followed a procedure of three stages: a first stage is geared towards identifying of main concerns about future developments; a second stage focuses on the discussion of key uncertainties and driving forces; and a third stage develops the actual scenarios. It is important to mention that this approach of scenario development was introduced for the first time to the majority of team members.

The workshops were initially based on a diversity of approaches to embrace different perspectives in a knowledge co-creation process. Through dialogue and storytelling, we aimed to shape the definition of phenomena and problems related to the future of Burkinabè fisheries. When starting storyline development several steps were taken within SUSFISH to increase the number of iterations, the most important being to start with an existing set of scenarios. This eliminated the most time-consuming step of building storylines from scratch, thereby speeding up the process and increasing the number of iterations of the Story-And-Simulation cycle. To that end, we collected this first set of scenarios with short narrations already prior to the workshop, which could serve us to elicit a set of important issues. With the purpose to elicit standardized narrations and to establish a common language across the project team to facilitate learning, we used a questionnaire, which was sent to all team members.

The questionnaire was designed for eliciting knowledge by focusing attention through a set of questions. It asked for 1) processes or trends influencing fishery sustainability, 2) key words, 3) scenario showing how sustainability is influenced (3 sentences or less), 4) important factors involved in this scenario, 5) relations between factors that influence this scenario (factors involved, how they interact, results) and finally 6) key questions or uncertainties. The questionnaire was sent out via e-mail and team members had about three weeks to fill it and send it back before the workshop. Further information how to use the questionnaire was provided only written.

Except for the students in Austria, not attending the workshop in Burkina Faso, an extra workshop on storytelling and formulating key questions was held in advance. The response to this questionnaire from Burkinabè partners was very low, only a few questionnaires on issues in natural sciences came back, and therefore the elicitation of key parameters was very limited. The reasons why this instrument was not functioning remained unrevealed. However, based on our experience of the effectiveness of face-to-face dialogue we adapted such questionnaires into a structured interview format because the questionnaire was obviously not acknowledged as a tool by the whole team.

Thus, the schedule was slightly changed and narratives were developed dialogically during two consecutive workshops at the beginning of the modelling exercise. We collected interactively short narrations among all participants. Every member was given a large degree of freedom, in terms of how to frame and express with own words and in a familiar language, to develop his or her own scenarios or phenomena. In most cases this led to a set of first storylines, in which the starting points were recognizable. In this first step it became evident, that only by working together to answer the questionnaire could the participants become come comfortable and fully respond to the technical language of systems analysis used to frame the questions. It was familiar to the participants, but in different shapes. For instance, the notion of a *scenario* was very abstract in the beginning, it was used as a quite technical term. Therefore, we needed to establish a common ground of understanding how to use this terminology and logic as a tool for our participatory approach. As we were not fully aware of this lacking common understanding, the two workshops were divided in a session focusing on biophysical issues and another for the socio-political questions. It was an important learning effect to the whole group, that it seemed to be more familiar to the experts from natural sciences to formulate key arguments, develop scenarios in three sentences as well as to trigger events. As a consequence, we encouraged all team members to participate in both workshops actively regardless of disciplines and professional expertise. The joint work in crossing disciplinary boundaries helped to identify key factors by breaking out of a technical aid speak and to integrate our different ways of knowing.

During the workshops, collaboration improved as we acknowledged and renegotiated the workshops' purposes, explaining in the process the methodological approach of scenario development and its importance to our findings. This effort was done in several days when we followed consequently the stories brought up in the workshops. In order to tackle complexity of problems we assessed together two types of knowledge: Knowledge academically collected and evaluated and on the other hand co-created knowledge between the involved researchers and practitioners, that is contextualized in social interactions and local systems. This process was quite intensive, enriching but also exhausting sometimes. As previously mentioned, it was necessary to pursue the lively debate in several languages. In some cases, it was necessary to evoke them in languages such as Jula or Moore. And because of plurilingual communication situations it was not evident that all the information and knowledge could be grasped by interpreting after the sometimes, long-lasting discussions to integrate them in the scenario development in scientific English. Translation became key to understanding and making meaning in this context.

5.2.2. The debate is open: translational practices to negotiate meaning

During the workshops, we experienced translation practices as a valuable instrument to negotiate meaning in order to experiment with new forms of thought and action - socially creative strategies - in order to understand problems and complexity not only through responsibilities, competences, and disciplines. We used them as a tool for the experiment of translating diverse theoretical concepts into the specific research contexts of fisheries and water management in our research areas. To a certain extent, translating showed us how theoretical concepts were constructed. It helped us to explore ways on how to adopt theory to practice.

But before it is important to elaborate on the notion of translation we use in this context. Translational practices do not apply simple linguistic relations, but go beyond equivalence and a copying of the original. We understand them as a continuous process of transformation (Bachmann-Medick 2009). Translation is the work of social worlds and not individuals (Gal 2015) and does not establish equivalence (Sakai 2006). It also involves appropriation and representation. The realm of translating language and text thus opens up to include a wider horizon of cultural translation practices (Bachmann-Medick 2006). It became a fundamental category of analysis in order to meet the transcultural challenges of our research practice and the contentious field of transcultural and transdisciplinary encounters. It is not the textual notion of translation into simplistic metaphors of transmission. An additional, decisive quality of this concept of translation is that it is tied to everyday life and agency and not to a few persons with cultural expertise. In our scenario development sessions, which were linguistically and theoretically very complex, we used this technique for dealing with multi-layered differences and disjunctions in knowledge, perceptions as well as in discourses not only bound to diverse languages, but also to different learning traditions.

Furthermore, interpreting practices among the participants were necessary in these sessions as French, English, Moore and Jula were very important means of communication amongst the participants, but not mutually understood by everybody. We experienced this necessity of interpretation during the discussions on the concept of *subsistence*. The term came up in the elaboration of key parameters for the commodification of fish. During this process, we identified subsistence fishing as a key parameter for questions of how to manage fisheries sustainably in Burkina Faso especially with regard to water health, sustainable water use around the reservoir and issues of environmental pollution. Farmers, who fish during the dry season, act individually, and are not organized within an association or governmental management authority. They use the resource of water for irrigation and for fishing adjusted for their personal needs, and this practice is governed by a traditional management system of natural resources. Introducing commercial fishing to this system exposes a gap in management. In order to integrate this aspect in our scenarios, first of all it was necessary to define subsistence fishing as a measurable category in contrast to commercial fishing. Politicians introduced it to our discussion as a very technical concept, clear-cut and static. It was framed in the development paradigm as *to meet the basic needs of food security*. But the method of systems analysis required us to define relationships to other parameters and along disciplines and revealed the need to better understand different interpretations of subsistence. For instance, in natural sciences it was necessary to clarify if subsistence is measurable by an average earning per fishermen.

Later, the question of surplus came up: is it still considered as subsistence activity with purchases in the market? Social sciences assumed that the short-term planning for livelihood might be a significant characteristic of subsistence, too. In ecology, the concept includes a dimension of sustainable social practice as it can be seen to contrast with larger-scale, more intensive commercial practices. The conflicting overlaps between interpretations seemed first to block attempts to agree on a clear-cut definition as our debate widened to a new dimension of complexity, instead of narrowing down to measurability. But by re-inventing the phenomenon *subsistence* through translational mediation we gained more understanding and became aware of misunderstandings. In this case, the term served as an interaction point of reference. Contradictions and misunderstandings indicated where we should critically reflect assumptions and prejudices. Indeed, the mode of thinking itself also benefits from translational qualities, and limitation of thoughts becomes obvious while reflecting on processes of mediation. As translation is not a unilateral one-way activity, we started with mediation activities. We used several languages for this debate, although some parts of the conversation were not accessible for everybody present in the workshop. The shift to

French and Moore was important, because then the meaning and perception were shifted, and, most importantly, mutually transformed. For instance, we gained insight regarding the hegemonial aspects inherent to this concept, such as the implicit connotation with poverty, primitive societies and backwardness framed by *tradition*. We assume that theoretical terminology is not simply a given static entity, but is created in a continuous process through translation in the first place. It also questions notions of origin as well as concepts of authenticity: in our case the distant habitus of policy-making institutions became evident, who fail to take into account dimensions such as special skills and knowledge of local resource users.

While we did not end up the discussion with a new clear-cut definition, translational processes allowed us to grasp in much more differentiated ways the term subsistence as it is widely used, but the process of learning how to understand better was very rich. In this regard we experienced traditional fishing practices as un-stable categories, and as a place of active transition and of cultural production of *creolization* - such as the translational adaptation of elements of French (neo)colonial dominant rhetoric, or its assimilations.

For us it was an important process to break up clustered, blanket conceptions of "intercultural" difference into singular steps of translation through which acts of understanding, mediation can be revealed, and misunderstandings and communication blockages become acknowledged rather than obscured. It induced us to reflect on how our own culturally specific positions - although seeming very objective, are loaded with ideologies on marginalization and disadvantage.

These activities are not necessarily always smooth or successful or capable of "bridging gaps", but differentiation enriches analysis. Furthermore, it embodies the basic elements for a self-reflection of transdisciplinarity. Only by exceeding the current limits of explorations at the borders of the disciplines it was possible to better understand the zones of overlap between different disciplines as perhaps conflictual yet productive and readily negotiable zones of translation. But it is vital to this approach of mediation processes to be sensitive to translational qualities, which allow differentiation and which enrich rather than simplify meaning. It is indispensably indirect, mediated by a third party and should include the acknowledgement of disruptions, rejections, misunderstandings and conflicts that can occur in our research processes. Most importantly this is tied to the ideological and perilous role of the translator him/herself as a cultural and language broker (Jacquemet 2005), which has to be reflected in every transdisciplinary research process. Translators constitute knowledge and speech style and precisely such qualities of inconsistencies, obstacles and resistance are predominantly obscured from view in participatory approaches (Gal 2015). Understanding is a fundamental aspect of knowledge production processes, these become more complex when multilingual practices and translocal contextualization, reconceptualization and the translation of knowledge are involved (Langthaler et al. 2012). In our interactions we had to deal with internationally established knowledge in English scientific and often political registers, by translating them on our own we started to reflect our own positions and to revise our professional rationality. But as Gal (2015: 233) argues, a semiotic analysis of these components of (Latourian) translation would clarify the parallel ongoing processes of transformation and recreation.

Much knowledge was produced and also gained by means of translation and by questioning our own culturally specific positions, analytical concepts, and theoretical assumptions. The understanding of local concepts is therefore the result of an ongoing process whose point of reference is situated within the complex field of conflicts defined as socio-cultural peculiarities. Here it was very important to no longer see cultures as holistic and self-contained phenomena, with a common ground for their contexts of meaning. Discussions of cultural features exhibited the interwoven discourses of colonial and postcolonial thoughts. Translation is not conceived of as a strategy of simplification and of diminishing the issue's complexity, i.e. it is not a dichotomist way of handling cultural differences. We had to deal with the complexity of differences - and we did not only establish differences, but also considered their scope of negotiation by applying a strategy of interpretation and mediation.

Part of such a task of translation required the consideration of power hierarchies and asymmetries that were also evident in knowledge traditions. To us it was important to become aware of the highly diverse set of hierarchical systems we were involved in. For the SUSFISH team the concept of gender balance within our research agenda provided an insight to this multi-layered, interferential system of concepts and ideologies.

5.2.3. The debate on gender

We made significant progress in examining aggregate patterns of social interaction and their influence on natural resource management. By discussing our different, sometimes diverging perceptions and also by interpreting certain issues, we could even deepen our understanding of other perspectives on gender topics. But in the beginning, it was a real challenge for the group to accept a gender sensitive approach applied across all disciplines. As a condition for funding, it was introduced as a hegemonic concept into the SUSFISH research agenda. The team coped with this challenge by engaging a gender expert. And because of her expertise and experience, she started to involve the team members, despite their opposition, in gender workshops and trainings in order to introduce gender-sensitive indicators and tools for data collection, as already outlined in the previous section. But the majority of in-country project partners continued to be irritated, questioning the meaningfulness of such trainings for biophysical scientists. During the scenario development workshops the debate on gender issues became quite imbalanced, reinforcing obscure assumptions on both sides.

In one of the first workshops, held after two years of cooperation, very controversial positions ensued on the usefulness of a general application of this concept on our project. The debate revealed the very abstract nature of this concept, which policy makers had introduced in a very technical manner and prioritized by making its study contingent on funding. It was framed in clear-cut definitions of expert's register that subsumes basic rights in a citational practice (see Gal 2015). Therefore, in the beginning of our discussions, it was a highly contested issue because the whole gender approach was reduced to the numeric equivalence between men and women and the conflict revolved around feelings of injustice. It was not acceptable for everybody that, for instance, male students applying for grants within the project should be disadvantaged because of the low rate of female researchers in our team. Some experiences of involving female students in fieldwork activities were negative, because of their low engagement and a very modest interest for bio-physics in general. Questions of structural discrimination or social exclusion of female students in the context of education and academic career development were not raised until the discussion turned on the family situation of a female researcher of the team. Now the gender expert started renegotiating the concept and could explain to her colleagues why a gender strategy in the fisheries sector also needs to integrate female perspectives. She started to translate this very theoretical concept by re-contextualizing it to the female researchers' perspective. At the same time, the gender expert's role completely changed within the team. She had followed the discussion for a long time without any substantial objection to the debate dominated by her male colleagues. But on the dispute about equal access for female students to natural sciences she started to provoke contradiction in order to elicit hidden assumptions.

At the beginning, in our questionnaires and hierarchization exercises, categories such as *marginalized people, equitable development* and *participation* were used like exclusive items, serving an expert jargon within the hegemonic paradigm of development. For the process of renegotiating meaning, it was very important to have an interaction on this in the team. Together we started to examine categories and analytical concepts in terms of their translatability in the realm of gender issues, to open them up and create a transgressive common understanding for our research agenda in several languages. And again, the concept of language we use is not a concept of bounded entities clearly separable by names and categories, such as Moore, French or English, but a space to learn about meaning in between varieties, linguistic features and jargons (Makoni et al. 2007). Terms such as marginalization were discussed among the Moore-speaking group in order to explore matters of oppression and discrimination of women in rural areas. It helped to realize that our perceptions of rural work and household organisation are quite distant to the life world of fishing communities. As a consequence, research data and preliminary results were shared, discussed and evaluated in workshops with women associations. The debate on gender issues in bio-physics remained controversial, especially about correlations of processing fish practices and the diversity of fish species. We could not agree about the integration of gender-related categories in our interpretation of data on biodiversity and water health, but it was important that we confronted contradictions. Misunderstandings changed in terms of exceeding the current limits of explorations at the borders of the disciplines. It clearly helped us to explore new, more appropriate definitions, which were developed on a common ground of understanding. We shared a broader "cross-categorical translation" as a way of opening up this mechanism in a critical manner (Chakrabarty 2000). It was possible to better understand the zones of overlap between different disciplines as perhaps conflictual yet productive and readily negotiable zones of translation. As participants in transdisciplinary dialogues, we were designated to discover new interconnections between allegedly different dimensions of social exclusion. It helped us to understand better that it is crucial to transdisciplinary research to have a team of women and men, regardless of the disciplines and research areas.

5.3. Conclusion and main learnings

This project illustrates how a transdisciplinary research approach can reveal important cross-sectoral phenomena in sustainable fisheries in Burkina Faso. SUSFISH was founded by natural scientists and is based mainly on biophysical science. It explores the possibilities in analyzing fisheries in order to manage them sustainably. Therefore, the conceptualization integrated knowledge from social sciences, but in a very technical manner and only in the final phase of the project. A few workshops on modelling and synthesis exercises among scientists and policy makers provided room for discussion across disciplinary boundaries. More than 100 scientists, students, experts, fishermen, and others were directly involved in SUSFISH for a four-year period of cooperation. Their participation on a platform to bring their knowledge together happened only at the end of the project, organized as a final conference meeting. Due to very limited financial resources, the synthesis process to establish a joint foundation of knowledge could not include them all. The project would have benefitted had the local experts on all levels been more involved.

There were moments of applied participatory research as described by Haller et al. (2016), but participatory research was not conceptualized as a comprehensive approach. Moreover, participatory research moments were restricted to the following three levels: field research, training for local groups, and scenario development. It is important to note that the first two moments, data collection in the field and training of local groups, were not considered to be important interdisciplinary learning contexts. Hence, they were organized along criteria of group membership, time, and place - and strictly within disciplinary boundaries. It was assumed that the third step, the synthesis process, would provide local experts' knowledge to the joint exercises among the core team - transmitted by the responsible scientists - in order to develop scenarios together. We realized that this strategy to integrate knowledge step by step from a local to a more general view was theoretically sound, but only partly successful in research practice. Had there been an ethnographic approach, such as participant observation and qualitative interviewing as back-up to accompany the activities with local practitioners, important insights would have been taken more into consideration by the scientific experts. This would have had practical implications for every step taken regarding the design and application of questionnaires, policies, scientific knowledge and would have made the integration of local ecological knowledge possible.

Ideological assumptions on local knowledge and the role of experts hindered a broader recognition of local expertise as data. This should be thoroughly analyzed, not only presented as experience gained or as individual stories.

In our reflections regarding the process of synthesis happening at the third level of participatory research moments, we concluded that the effort to look thoroughly across disciplines can never be comprehensive in such a short time and with only a few meetings involving all experts. SUSFISH can thus be regarded as an exercise allowing for joint learning processes, which, in turn, can help extract the most important factors shaping ongoing questions for future shared research.

But beyond that we also experienced participatory research as taking place largely in controversial debates, full of inconsistencies, obstacles, and resistance. The participants' mutual work of collecting, analyzing and negotiating the meaning of information involved acknowledging these disruptions, rejections, misunderstandings, and conflicts. We used these meetings in a similar way as described for the constitutionality process (Haller and Merten 2018) and experienced them as a strategy for politicians and scientists to examine the dynamic implications of the facts and questions generated by the project. Very important was the process in terms of negotiating meaning and finding a compromise.

In retrospect, basic obstacles were the following:

- A short time frame: The transdisciplinary learning process was considerably more intensive as anticipated and would thus have needed much more time for discussion, as it took place largely through controversial debates. Therefore, transdisciplinary research activities are one of the most important contact zones, which should be planned rigorously from the beginning. An important lesson learned is to organize research activities along a continuous sequence of dialogic transdisciplinary activities for the whole duration of the project.
- One core element of our SUSFISH research was to establish and intensify cooperation among the many partners (nationally in Burkina Faso and in Austria as well as internationally between institutions of both countries involved) and was obviously too ambitious. The organization of a set of cooperation-promoting activities, such as joint interdisciplinary field trips, transdisciplinary workshops, joint lectures, shared supervising models for students (cross institutionally), system analysis, and concept modelling, was thus a challenge as well.

- Power asymmetries regarding class and gender: We realized too late that integrating students and important stakeholders in these learning and evaluation processes would have been very important. But a key question is how to organize such a space for debate in terms of time, languages, and power hierarchies. Another challenge was how to provide platforms to engage different stakeholders involved in the fisheries management at all levels and in group compositions, thus allowing them to interact.
- Power asymmetries are interwoven with language issues as well. An additional challenge is the issue of access in terms of language, education, and knowledge in such participatory debates. The involvement of diverse linguistic resources implies a reflection on prejudices, ideologies, and an understanding of roles. It also requires methods of linguistic anthropology to reflect on the exertion of symbolic power and inherent power relations the team was not aware of.
- Much knowledge was produced and also gained by means of translation and by questioning our own culturally specific positions, analytical concepts, and theoretical assumptions. The process of translation was more than the transfer of knowledge from one word to another. But we did not assess these communicative practices qualitatively. Later on, it was not possible anymore to analyze our social interaction in detail. The approach definitely requires ethnographic tools in order to grasp the process of knowledge framing and codifying, the translational practices within our disciplines, and to learn how others do it.
- Another obstacle to overcoming scientifically-based language we experienced in the SUSFISH project is written into the paradigm of development cooperation. This approach by its very nature implies inherent power asymmetries and imbalanced relationships due to obscure assumptions and expectations. These were implicit and therefore remained uncontested for a long time. But the conceptual space of modelling workshops obliged us to formulate some of our concerns and ideologically loaded assumptions. In addition, our discussions revealed that language and terminology play a crucial role in reconfirming development-driven concepts as well. Scientifically codified language was removed step-by-step for systems analysis reasons in order to open up a discussion about prejudices, ideologies, and an understanding of roles. We experienced plurilingual practices as a very important tool to allow for a dialogue on a par with each other.

Studying fisheries with natural science techniques based primarily on biophysical research does, in itself, not produce the requested solutions for sustainable management as our findings in social, economic, and political sciences showed. The exercises in systems analysis revealed a lack of insight regarding the patterns of social systems (blind spots), which are very much related to the ecological systems. The results of case studies in various sites clearly showed the need to understand local economies, political structures, and plurilingual practices as well as local knowledge. Yet even more prominent was the need to understand existing, often inherent power relations among local groups as well as among international experts. This experience includes a deeper understanding and learning in a transdisciplinary process in all research activities. A recurrent challenge to this method was the planning and working in a broad heterogeneous team.

5.4. References

Bachmann-Medick, D. (2006). Cultural Turns. Neuorientierungen in den Kulturwissenschaften. Reinbek bei Hamburg: Rowohlt.

Bachmann-Medick, D. (2009). Introduction: The translational turn. Translation Studies, 2(1), 2–16.

Berkes, F. (1999). Sacred ecology. Traditional ecological knowledge and resource management. Philadelphia: Taylor and Francis.

Bhabha, H. (2000). Die Verortung der Kultur. Tübingen: Stauffenburg.

Bourdieu, P. (1996). Understanding. Theory, Culture and Society: Explorations in Critical Social Science, 13(2), 17–37.

Chabwela, H.N. and Haller, T. (2010). Governance issues, potentials and failures of participative collective action in the Kafue Flats, Zambia. International Journal of the Commons, 4(2), 621–642.

Chakrabarty, Dipesh. (2000). Provincializing Europe. Postcolonial thought and historical difference. Princeton: Princeton University Press.

CNID-B. (Comité National d'Irrigation et de Drainage du Burkina Faso). (2010a). Diagnostic participatif du périmètre irrigué de Karfiguéla. Internal report. Ouagadougou, Burkina Faso: IWMI.

CNID-B. (Comité National d'Irrigation et de Drainage du Burkina Faso). (2010b). Diagnostic participatif du périmètre de Talembika. Internal report. Ouagadougou, Burkina Faso: IWMI.

Foucault, M. (1981). The order of discourse. In: Young, R. (ed.). Untying the text. A post-structuralist reader (pp. 48–78). New York: Routledge and Kegan Paul.

Folke, C., Hahn, T., Olsson, P. and Norberg, J. (2005). Adaptive governance of social-ecological systems. Annual Review of Environment and Resources, 30, 441–473.

Gal, S. (2015). Politics of translation. Annual Review of Anthropology, 44, 225–240.

Haller, T., Acciaioli, G. and S. Rist. (2016). Constitutionality: Conditions for crafting local ownership of institution-building processes. Society and Natural Resources, 29(1), 68–87.

Haller, T. and Merten, S. (2008). "We are Zambians—Don't tell us how to fish!" Institutional change, power relations and conflicts in the Kafue Flats fisheries in Zambia. Human Ecology, 36(5), 699–715.

Haller, T. and Merten S. (2010). "We had cattle and did not fish and hunt anyhow!" Institutional change and contested commons in the Kafue Flats floodplain (Zambia). In: Haller, T. (ed.). Disputing the floodplains (pp 301–360). African Social Studies Series 22. Leiden: Brill.

Haller, T. and Merten, S. (2018). Crafting our own rules: Constitutionality as a bottom-up approach for the development of by-laws in Zambia. Human Ecology, 46(1), 3–13.

Jacquemet, M. (2005). Transidiomatic practices: Language and power in the age of globalization. Language and Communication, 25, 257–277.

Kaboré, I., Moog, O., Alp, M., Guenda, W., Koblinger, T., Mano, K., Ouéda, A., Ouedraogo, R., Trauner, D. and Melcher, A. (2016a). Using macro-invertebrates for ecosystem health assessment in semi-arid streams of Burkina Faso. Hydrobiologia, 766(1), 57–74.

Kaboré, I., Jäch, M.A., Ouéda, A., Moog, O., Guenda, W. and Melcher, A. (2016b). Dytiscidae, Noteridae and Hydrophilidae of semi-arid waterbodies in Burkina Faso: species inventory, diversity and ecological notes. Journal of Biodiversity and Environmental Science, 8(4), 1–14.

Kok, K., Van Vliet, M., Dubel, A., Sendzimir, J. and Bärlund, I. (2011). Combining participative backcasting and explorative scenario development: Experiences from the SCENES project. Technological Forecasting and Social Change, 78, 835–851.

Langthaler, M., Witjes, N. and Slezak G. (2012). A critical reflection on knowledge hierarchies, language and development. Multicultural Education and Technology Journal, 6(4), 235–247.

Mano, K. (2016). Fish assemblages and fish-based assessment of the ecological integrity of river networks in Burkina Faso. Unpublished PhD dissertation. BOKU, University of Natural Resources and Life Sciences, Vienna.

Mahé, G., Paturel, J.E., Servat, E., Conway, D. and Dezetter, A. (2005). Impact of land use change on soil water holding capacity and river modelling of the Nakambé river in Burkina-Faso. Journal of Hydrology, 300, 33–43.

MAHRH (Ministère de l'Agriculture, de l'Hydraulique et des Ressources Halieutiques). (2003). Action Plan for Water Resources Integrated Management (PAGIRE). Ouagadougou. Burkina Faso: MAHRH.

MAHRH. (2006). Politique nationale de développement durable de l'agriculture irriguée. Stratégie, plan d'action, plan d'investissement à l'horizon 2015 – Rapport principal, MAHRH, Ouagadougou, Burkina Faso: MAHRH.

Makoni, S. and Pennycook, A. (eds.). (2007). Disinventing and reconstituting languages. Clevedon: Multilingual Matters.

Melcher A., Ouedraogo, R. and Schmutz, S. (2012). Spatial and seasonal fish community patterns in impacted and protected semi-arid rivers of Burkina Faso. Ecological Engineering, 48, 117–129.

Melcher A., Ouedraogo, R., Moog, O., Slezak, G., Savadogo, M. and Sendzimir, J. (2018). Healthy fisheries sustain society and ecology in Burkina Faso. In: Schmutz, S. (ed.). (2018). Riverine ecosystems management. Science for governing towards a sustainable future (pp. 519–540). Aquatic Ecology Series 8. Cham: Springer.

Ouédraogo, R. (2010). Fish and fisheries prospective in arid inland waters of Burkina Faso, West Africa. Unpublished PhD dissertation. BOKU, University of Natural Resources and Life Sciences, Vienna.

Peterson, G.D., Cumming, G.S. and Carpenter, S.R. (2003). Scenario planning: a tool for conservation in an uncertain world. Conservation Biology, 17, 358–366.

Petit, O. and Baron, C. (2009). Integrated Water Resources Management: From general principles to its implementation by the state. The case of Burkina Faso. Natural Resources Forum, 33, 49–59.

Sally, H., Lévite, H. and Cour, J. (2011). Local water management of small reservoirs: Lessons from two case studies in Burkina Faso. Water Alternatives, 4(3), 365–382.

Sakai, N. (2006). Translation. Theory Culture Society, 23(2–3), 71–86.

Sanon, V.-P. and Toe, P. (2015). Gouvernance et institutions traditionnelles dans les pêcheries de l'ouest du Burkina Faso. Etudes africaines. Paris: L'Harmattan.

Sendzimir, J., Reij, C.P. and Magnuszewski, P. (2011). Rebuilding resilience in the Sahel: regreening in the Maradi and Zinder regions of Niger. Ecology and Society, 16(3), 1.

Sendzimir, J., Magnuszewski, P. and Gunderson, L. (2017). Adaptive management of riverine socio-ecological systems. In: Schmutz, S. and Sendzimir, J. (eds.). Riverine Ecosystem Management - Science for governing towards a sustainable future (pp. 301–324). Aquatic Ecology Series 8. Cham: Springer.

Stranzl, S. (2014). Quantification of human impacts on fish assemblages in the Upper Volta catchment, Burkina Faso. Unpublished Master thesis. BOKU, University of Natural Resources and Life Sciences, Vienna.

Thompson Klein, J. (2004). Prospects for transdisciplinarity. Futures, 36(4), 515–526.

Venot, J.P. and Cecchi, P. (2011). Valeurs d'usage ou performances techniques: Comment apprécier le rôle des petits barrages en Afrique subsaharienne? Cahiers Agricultures, 20(1-2), 112–117.

6. Conclusion
Explorations and lessons for shared research

Claudia Zingerli and Tobias Haller

6.1. Explorations

This book is about researching and interpreting African environments by reading the landscape and the use of natural resources through different scientific, societal and political lenses. Inevitably, this creates struggles between disciplinary research traditions and emphases, and between the interpretations of the researchers and those of the researched. None of the authors in this book shied away from these struggles. They made productive use of natural science data, treating this information as being of equal importance to the practices and rationales of natural resource users.

In our opening chapter we described a broad spectrum of research in African environments, and demonstrated this breadth by reference to leading scholars who combined generalist and specialist knowledge about human-environment relationships. Our explorations went far back in time - something that in the business of science in the twenty-first century is rarely possible. They referred to the paradigmatic work of Fairhead and Leach (1996) and Berkes (1999) on African environments, institutions and the role of conservation and natural resource management for humanity and nature in the late twentieth century, and they went back even further in time to the nineteenth century, reflecting on Alexander von Humboldt's approaches to researching America's environments (see Wulf 2016). These explorations emphasized the merits not only of a mixed methodology but of the turn towards integration in research on (African) environments (see Bornemann et al. 2017), combining disciplinary methodologies as well as the essentials for knowledge production based on the relationships between researchers and researched.

Now, in our final chapter, we bring into focus multidisciplinary and mixed methods research and their methodological and learning implications. The articles collected in this edited volume speak of socially and environmentally grounded research that requires ample time and familiarity with specific contexts at various levels, enabling learning as a multidimensional and multilevel process. They reveal key turning points in the participatory research process that arise from collaboration, exchange and intense debates about the role of language and translation in interdisciplinary and intercultural research settings. All the articles have a distinctive approach to the researcher-researched nexus, and are aware of the emergent character of such interactions.

For our final explorations regarding research processes, we again go back in time. We have found it useful to analyse how the contributions of Prudat et al., Oyama, Jewitt et al. and Slezak et al. mirror the basic principles of action anthropology, which was initiated as early as the 1950s (Tax 1975). Sol Tax's initiative in the 1950s was grounded on the understanding that working with the first nation indigenous peoples of the Meskwaki (also called Fox Indians) in their remaining small reserves only made sense if this work was co-research on topics relevant to these people. Moreover, the insights into these topics would be linked to actions in which the anthropologists became advocates for these people in domains in which the local groups did not have the knowledge or power to act. Tax labelled this process action anthropology (see Tax 1975), and his approach was later used by a few anthropologists in the 1980s and 1990s (see Schlesier 1980, Seithel 2000, 2004), but regained momentum in a special issue by Stapp et al. (2012). Foley (1999) and Bennett (1996) then published several critical reflections on how this, often unrecognized, approach had influenced the more widely accepted applied anthropology approach, in which local involvement in research, especially on development issues, was created. In contrast to the action anthropology approach of cocreation and concerted action, applied anthropology is usually implemented as an anthropology *for* but not necessary *with* local people. Applied anthropology tries, to solve practical problems in many other fields as defined only by local people, such as health and medicine, business, education, environmental issues, community development, disaster research and international development (Van Willigen 2002). Applied anthropology, however, tends to lack the critical means to integrate, appreciate or anticipate local knowledge, in contrast to action anthropology, whose creators put at centre stage the local people *and* the researchers.

In the four articles collected in this book, elements of action and applied anthropology can be detected, whether or not the researchers are trained anthropologists. All the researchers demonstrated an openness towards different disciplines and scientific domains, and exposed themselves to situations of uncertainty in researcher-researched relationships. In all cases, living through discomfort (such as translation challenges and conceptual discrepancies), unexpected and counterintuitive actions led them to come up with accounts that speak about social learning and the co-creation of knowledge, enabling a critical reflection on how research in African environments can evolve. In this way they combined different disciplinary research traditions with a more rigorous and extended approach, taking into account local perceptions about issues and then readjusting to take their next steps.

In the following we look at four elements that seem particularly striking for a reflection about what we would like to call "shared research", by which we mean research co-produced for making use of the fluidity of knowledge about natural and social processes in African environments, in order to demonstrate that there are multiple ways of dealing with the challenges of unsustainable development. These four elements are:

- Learning as a multidimensional and multilevel process over an extended time and scale;
- The dimensions of participatory research;
- The role of language and translation in interdisciplinary and intercultural research settings; and
- Turning points in collaborative research processes.

We also argue that these processes represent journeys that are closer to action anthropology than to applied anthropology, and that this position is needed in order to understand local issues and to engage in a process of shared and also co-owned research.

6.2. Learning as a multidimensional and multilevel process

The literature distinguishes between different orders of learning. According to Sterling (2011), first order learning refers to the reproduction of knowledge and "doing things better", and second order learning to critical reflection and "doing better things". In addition to these two orders, a transformative form of learning exists, leading to an experience of reflecting about our worldview rather than seeing with our worldview. Such a reflection enables us to be more open to, and to draw upon, other views and possibilities (Sterling 2011).

According to Land et al. (2014), transformative learning implies a paradigm change triggered by experiences of liminality. This paradigm change is an in-between state of ambiguity or disorientation in the process of learning and understanding, and has effects on knowledge production processes. We would also argue that a shared learning process, in which all partners in the process are mutually perceived as providing knowledge, creates the dignity and respect that are the basis for learning collaborations. Like the Fox project in which Sol Tax and his students became engaged with the local first nation people (Foley 1999) and, in another context, research on common pool resource management in African contexts (Haller 2010), the initial top-down approach in each of the processes portrayed gave way to a sharing of knowledge with local actors, and co-developing and enabling actions.

All four contributions in this edited volume are framed by reference points given by global or regional policy frameworks, such as international soil categorization and the implications of soil resource management, policies for afforestation and combatting desertification in the Sahel, efficient energy use by promoting the use of cookstoves, or the multiple use of water reserves for both fish and energy production to enhance food security. The researchers engage in one way or another with understanding institutional design and the governance of common-pool resources in a global-local world (see also Haller et al. 2019). They all contextualize the policy frameworks in which the research is embedded, using thick descriptions of how those frameworks play out in their research processes and in various contexts and localities - and they reveal how much those frameworks differ from the assumptions and conceptualizations of the African environments. All four contributions reflect on their stages of ambiguity or disorientation, Prudat et al. by juxtaposing local knowledge and international soil classifications, Oyama by liminal experiments using urban waste for improving soil fertility and land management, Jewitt et al. by debating cookstoves in culturally and economically diverse settings, and Slezak et al. by addressing the limits of inter- and transdisciplinary research when mind-sets and values are unintentionally reiterated in research collaborations.

Oyama demonstrates, in his account based on extended and experimental field research in Niger, what it means to learn and unlearn in situations of increasing environmental, socio-economic and political stress. He is able to do this because for more than 15 years he returned to his research sites; he is a learner, as well as an advocate for co-creating, with local authorities and leaders, solutions to truly wicked problems that go far beyond his initial focus on the material basis of soil and land resources in the Sahel. His account shows that the learning at the local level emerges in a completely different way from what was intended by the Greening Sahel policy, and that policies that do not address the systemic complex of the environmental, social and political dimensions of the problem can reinforce or spur on new conflicts. This approach of co-research and learning also speaks about managing common pool-resources in a new and participatory way, and includes elements of action anthropology and constitutionality. It is about trying to understand local solutions and then involving local actors in the common research agenda; this then leads to collectively defined actions, in which the local actors can also see themselves as owners of the process. As Oyama provided a neutral platform for interaction, this constellation was also enabled in a way that resonates well with the constitutionality approach (see Haller et al. 2016, 2018).

The learning process in the article by Jewitt et al. is characterized by multi-stakeholder perspectives on specific events used for experimenting with the clean cookstove technology developed for the global south in the United Kingdom (UK). Their account challenges the promotion of cookstoves and their potential as cost-effective clean-fuel solutions, as propagated by global policy drivers. The researchers' experiences in the UK helped them to design a methodology for fieldwork and to compare experiments in different localities with a view to upscaling them to enhance policy frameworks. They applied, in their learning and knowledge production process, a fluid and varied use of technology, taking into account socio-economic contexts and real-life situations in which the cleaner cookstoves were just an option but not the only solution to the problem of fuel and energy shortages. Knowledge about materiality and practicability was gathered in a way that took local views and knowledge seriously and that also created notions of co-ownership for this process.

By doing participatory and interdisciplinary research with joint data collection, Slezak et al. created learning environments that included social and cultural aspects as well as institutional and legal frameworks. They show how water and fisheries management in water storage facilities in Burkina Faso provided the basis for a more participatory research project called SUSFISH (Sustainable Management of Water and Fish Resources). The basis of this project was an interdisciplinary approach, which recognized the failure of previous projects that were predominantly based on technical expertise. With their transdisciplinary approach, they gathered social, economic and political information to create pathways for more sustainable fisheries in

Burkina Faso. The very set-up of their project provided uncountable multidimensional and multilevel learning processes. These are explored in more detail in the following sections.

The specific collection of these four contributions leads us to conclude that the dimensions of learning are expanded if there is room for critical reflection on research processes and multiple expectations. By formulating (in writing or speech) deeper insights and discomforts, changes of perspective on the research process can shift paradigms and epistemological traditions.

6.3. Dimensions of participatory research

Being aware of different orders of learning can also contribute to an enhanced understanding of the dimensions of participatory research in African environments. Key elements of participatory research are sequential reflection and action. Participatory research is carried out with and by local people rather than on them. The key difference between participatory and conventional methodologies lies in the location of power in the research process (see Cornwall and Jewkes 1995), which speaks to key elements of action anthropology and constitutionality.

The four contributions in this book all have different approaches to participatory research. Two of them were initiated by research consortia at Swiss and German (Prudat et al.) and Japanese (Oyama) universities. Both of these groups chose the approach of spending longer time on fieldwork in, respectively, Namibia and Niger, during which they collaborated closely with local people as well as with interpreters. For Prudat, who was trained as a natural scientist, the collaboration with his local research assistant was key to any form of data collection and understanding of local people as well as socio-economic circumstances. Oyama, who was trained as environmental scientist, eventually chose the classical anthropological approach of participatory observation, which enabled him to establish relationships with local authorities and leaders and to obtain access to local communities and their livelihoods, as well as their increasing limitations.

The two other contributions were initiated by mixed consortia composed of UK- and Austrian-based researchers as well as researchers from Burkina Faso, Nigeria, Kenya and Malawi. They used their contextual and embodied knowledge to make use of the transdisciplinary research methodology they developed. Both groups debated their methodology in mixed teams of researchers and informants of different backgrounds, sometimes in places that were far away from the actual application of the cookstove technologies or the fisheries management scheme. The combination of the local knowledge of researchers from the research context (Benu state) and the involvement of researchers with links to the administration in Burkinabè fisheries provided a setting for structuring the research process using the shared capacities of local as well as distant knowledge.

The four contributions lead us to conclude, in our reflection on "shared research", that carrying out participatory and integrative research in African environments means being explicit about the roles of those who move in and out of the local context and those who stay. Extended stays in local contexts and the sharing of daily life experiences help to develop a sense of diverse power relations in research teams as well as in researcher-researched relationships. Also, perseverance during challenging workshops and exchanges supports the sequential reflection and action that are typical participatory research dimensions. Key elements for more inclusive learning and better research ethics are mutual respect and trust between unequal research partners and those who are researched. Such research processes require careful planning and an openness towards emerging participation while the research progresses. That also means to accept limitations and to cope with frustration as participatory moments can become overly complex.

6.4. Role of language and translation in interdisciplinary and intercultural research settings

Multidimensional learning and participatory research inevitably touch on different understandings and epistemologies, and their various expressions in language and speech. In the participatory research set-ups described in the contributions in this edited volume, it was inevitable that language and translation had to be addressed and worked through. These can again represent a sort of liminal experience (see Land et al. 2014), as they challenge worldviews and multiply the possibilities of interpretation. The translation can also act as a way to manage the states of ambiguity or disorientation in the processes of data collection, analysis and interpretation.

There was a moment of discomfort for Prudat et al. when it became clear that the explicit language used in international soil classification neither did justice to nor made sense in the context of nuanced and embodied (through manual labour) knowledge of soils and their fertility. Prudat and his research assistant created a way to reduce the complexity by developing a sort of code in the translation process. They also deliberately did not deal with the data that could not be fully deciphered and thus could not be linked to their research questions. They opted for data and interpretations that built bridges between local knowledge, which was relevant for livelihoods and agricultural production, and the international soil classification. In this way they could link and contextualize the local knowledge with policy measures and natural resource management regulations. The discomfort of potentially losing out on novelty and understanding because of decisions related to language and interpretation is impressively described in their account.

The translation between engineers and social scientists as well as between researchers and cookstove users locally and in the UK was a key element of Jewitt et al.'s experimental set-up. They designed and used bake/cook-off events as boundary spaces for enhanced understandings of cookstove use. The translation and interpretation in the various contexts in which the bake/cook-off events were subsequently organized and in the testing of the use of cookstoves in prolonged field research were key for their insights into the understanding of the use and rejection of improved cookstoves. The different contexts cross-fertilized their understanding and contributed to a greater robustness.

Translation was a central element for the nine tools of cooperation that were used by Slezak et al. Translation spanned joint data collection, the integration of local knowledge, and gender-sensitive workshops and public conferences with practitioners, decision makers and scientists. However, the project had a natural scientist language orientation and, at the beginning, did not reach a level at which discourses on findings could be translated into other disciplines or into other social and cultural contexts. Thus, the impression emerged that only natural science language mattered, and this was challenged at the many workshops in which a more common type of language was an issue. This meant that discussions framing the concepts by using natural sciences would not take the project far enough, and it became evident that social sciences, as well as local knowledge, were key elements that needed to be incorporated.

With respect to language and translation, we conclude that openness in the research process for corrections and greater inclusion characterizes the approach for "sharing" in collaborative and integrative research on African environments.

6.5. Turning points in collaborative research processes

Finally, the liminality of learning experiences, as expressed in what we call distinct turning points, is a characteristic of the specific contributions that make up this edited volume. In each of the four contributions it is possible to detect such moments of experienced change. Living through these allowed the researchers to continue to, and to complete, the next stage of the project. In Prudat et al.'s case the turning point happened between the researcher and the local assistant who jointly found a way to translate, interpret and, finally, to make sense of rather controversial data and insights. The turning point in Jewitt et al.'s account was the first bake/cook-off event that was used for the systematic collection and understanding of diverse preferences, making use of this boundary space for including many perspectives and opening up space for interpretation.

In Oyama's case, we detect a turning point after he consulted with the village leaders and started partnering with local people to hire them for work. In his writing, his language changes from "I" to "we". Together, they started to develop and test a solution to a wicked problem, that of enhancing soil fertility and fodder production, with a potentially mitigating effect on smouldering conflicts about land. Together they seem to have been united and determined to create a situation that might enable the local people to deal with the fragility and temporarily reduce the environmental and social pressure.

Several turning points can be detected in the account by Slezak et al. While the project was set up as a senior-level transdisciplinary research partnership, it was the group of students who managed to include local knowledge in the overly expert (natural) scientific notions of the problems. The group of students expressed an openness towards local fishery experts in the community, and started to interact directly for longer periods of time with local fishermen, enabling them to shape and influence the research practice by direct communication and exchange. Like Oyama's turning point, a joint fishing technique was developed as a consequence of this. The joint technique allowed for a comparison of catch issues at the various sites. This included not only word for word translation but also the translation of concepts concerning water and fisheries.

In a later stage of Slezak et al.'s project, another turning point was marked by the challenges emerging with respect to gender, which became important during the different workshops. At some point it became clear that the procedures in the workshops excluded women from the discussion processes, and implied that their knowledge mattered much less than men's knowledge framed in development language. This created space for more nuanced reflections on the inter- and transdisciplinary setting of the project. However, it required facilitators who were open-minded and bold to address unequal power and gender relations in the research process, which had been structured as participatory and collaborative but was shaped by specific mind-sets and epistemological traditions.

To conclude, unplanned and unanticipated chances and challenges emerge in collaborative and integrative research. Being open to and respectful of the diversity of actors, both researchers and researched, and of their perceptions and contributions, can enhance learning and the facilitation of a common understanding of key aspects in researching African environments. However, as with the action anthropology and constitutionality approaches, these cases show a central finding: being open to the unexpected and learning from these experiences as well as reducing the power of white, male, northern-based researchers in scientific contexts is a key element of moving towards shared research. Only when this power asymmetry is reduced might research become more participatory, and this participatory research will also lead to better scientific outputs, as it profits from more detailed knowledge and is a better basis for action.

6.6. Towards shared research

This edited volume was produced with the intention of drawing attention to the twists and turns that emerge in intercultural, interdisciplinary and transdisciplinary, participatory and integrative research in African environments.

Our call for "shared research" emphasizes an openness to use diverse perspectives and not to shy away from the complications and complexities of local knowledge and development contexts. It should encourage and motivate the creation of direct encounters between researchers and researched in various contexts for production processes of joint knowledge, combining various concepts of management and development. We conclude with three final suggestions:

- *Making room for long-term research engagement with extended fieldwork stays in local and regional contexts*: Long-term research engagement enables local voices to be heard and understood. At the same time, it enables local actors to be in a position to understand the views of external researchers and experts. This happens by the sharing of everyday activities and living conditions, and is supported by applying a participant observation research methodology. Long-term research engagements can create trust, a key element for exchanging information with each other and knowing that all parties are trying to understand each other. For researchers, this requires an openness to local ways of doing and seeing things. One researcher who adopted this principle with complete success was none other than Alexander von Humboldt, who, while being extremely interested in natural scientific data, always showed an interest in local views and rationales. To him, these were as important as the European western scientific views he represented (see Eibach 2012, Wulf 2016). In addition, contextualization matters.

- ***Contextualizing research projects, by referring both to diverse scientific contributions (including those published more than five years ago) and to the global drivers that shape development and livelihood contexts today***: Without a concise contextualization of legal and power-specific issues there is little room for collaborative research. Trying to understand not what reality is in singular terms but what the elements of the different views on realities are is a crucial step in participatory and integrative research in African environments. Including local views at the same level as scientific language and knowledge leads to more robustness in the research process and a better preparation for outreach and implementation of the research results. Such shared research evolves from "they do it" to "we do it". Obviously, there are risks: such processes are emergent and often unpredictable, and they can suffer from drawbacks such as those represented in gendered patterns of knowing or epistemological dominations. Contextualizing thus also means providing spaces and platforms for direct and open discussion and constructive, inclusive debate.

- ***Making research processes and methodological challenges more explicit***: Last but not least, we set out one of the important lessons of this collection of articles. "Shared research" is not a one-way street, but is full of twists and turns as well as conflicts. An analysis of the processes that are occurring, and speaking about and discussing where the team, with its different parts and functions, stands are of central importance in finding ways for continuing the shared research process. Being more explicit about the research process and the methodological challenges of research endeavours is a way to give justice to the multiple learning loops and the emergent character of results in researching the wicked problems of today (see KFPE 2018).

All these elements echo the beginnings of action anthropology and constitutionality processes, and indicate that "shared research" can be a starting point but must also be a self-reflective process that should anticipate the different interests and power relations of all the stakeholders. The challenge of "shared research" is to keep the process running in a participatory way and to mitigate power asymmetries. It also enables mistakes to be made but gives the capacity to learn from and be creative about them. This edited book tries to provide an input for this type of co-research and learning.

6.7. References

Berkes, F. (1999). Sacred ecology. Traditional ecological knowledge, and resource management. Philadelphia, PA: Taylor and Francis.

Bennett, J. W. (1996). Applied and action anthropology: Ideological and conceptual aspects. Current Anthropology, 37(1) Supplement: Special Issue: Anthropology in Public, 23–53.

Bornemann, B., Bernasconi, A., Ejderyan, O., Schmid, F., Wäger, P., Zingerli, C. (2017). Research on natural resources: The quest for integration revisited. GAIA, 26(1), 16-21.

Cornwall, A. and Jewkes, R. (1995). What is participatory research? Social Science and Medicine, 41(12), 1667–1676.

Eibach, J. (2012). Tasten und Testen. Alexander von Humboldt im Urwald. zeitenblicke, 11(1).

Fairhead, J. and Leach, M. (1996). Misreading the African landscape: Society and ecology in a forest-savanna mosaic. Cambridge, UK: Cambridge University Press.

Foley, D. E. (1999). The Fox project: A reappraisal. Current Anthropology, 40(2), 171–192.

Haller, T. (ed.). (2010). Disputing the Floodplains: Institutional Change and the Politics of Resource Management in African Floodplains. African Social Studies Series 22. Leiden: Brill.

Haller, T., Acciaioli, G. and Rist, S. (2016). Constitutionality: Conditions for crafting local ownership of institution-building processes. Society and Natural Resources, 29(1), 68–87.

Haller, T., Belsky, J. M. and Rist, S. (2018). The constitutionality approach: Conditions, opportunities, and challenges for bottom-up institution building. Human Ecology, 46(1), 1–2.

Haller, T. Breu, T., de Moor, T. Rohr, C. and Znoj, H. (2019). Introduction on commons in a glocal world: linking local and global systems, power processes and local reactions in the management of common pool resources. In: Haller, T., Breu, T., de Moor, T., Rohr, C., and Znoj, H. (eds.). The Commons in a Glocal World Global Connections and Local Responses (pp. 1–20). London: Routledge.

Haller, T. and Merten, S. (2018). Crafting our own rules: Constitutionality as a bottom-up approach for the development of by-laws in Zambia. Human Ecology, 46(1), 3–13.

KFPE (Swiss Commission for Research Partnerships with Developing Countries). 2018. A guide for transboundary research partnerships: 11 principles and 7 questions. 3rd edition. Bern: SCNAT.

Land, R., Rattray, J. and Vivian, P. (2014). Learning in the liminal space: A semiotic approach to threshold concepts. Higher Education, 67(2), 199–217.

Schlesier, K. H. (1980). Zum Weltbild einer neuen Kulturanthropologie. Erkenntnis und Praxis: Die Rolle der Action Anthropology. Vier Beispiele. Zeitschrift für Ethnologie, 105(1), 32–66.

Seithel, F. (2000). Von der Kolonialethnologie zur Advocacy Anthropology. Hamburg, Münster: Lit Verlag.

Seithel, F. (2004). Advocacy anthropology: History and concepts. Revista Anthropológicas, 15(1), 5–48.

Stapp, D. C. (ed.) (2012). Action anthropology. Sol Tax in 2012. The Final Word? Richland, WA: Northwest Anthropology LLC.

Sterling, S. (2011). Transformative learning and sustainability: Sketching the conceptual ground. Learning and Teaching in Higher Education, 5, 17–33.

Tax, S. (1975). Action anthropology. Current Anthropology, 16(4), 514–517.

Van Willigen, J. (2002). Applied anthropology. An introduction. South Hadley, MA: Bergin und Garvey.

Wulf, A. (2016). Alexander von Humboldt und die Erfindung der Natur. Munich: Random House GmbH.

Authors

Peter Atagher obtained his PhD from the Faculty of Engineering, University of Nottingham, United Kingdom.

Lena Bloemertz is a senior researcher at the Centre for African Studies and the Department of Environmental Sciences of the University of Basel, Switzerland.

Mike Clifford is Associate Professor at the Faculty of Engineering, University of Nottingham, United Kingdom.

Olivier Graefe is professor of Human Geography at the University of Fribourg, Switzerland.

Sarah Jewitt is Professor of Human Geography and Development at the School of Geography, University of Nottingham, United Kingdom.

Tobias Haller is professor at the Institute of Social Anthropology of the University of Bern, Switzerland.

Colette Kabore is collaborator at the General Directorate for Fish Resources, Ministry of Animal and Fish Resources Ouagadougou, Burkina Faso.

Nikolaus Kuhn is professor of Physical Geography and Environmental Change at the University of Basel, Switzerland.

Andreas Melcher is senior scientist at the Centre for Development Research, BOKU University of Natural Resources and Life Science, Vienna, Austria.

Paul Meulenbroek is scientific collaborator at the Institute of Hydrobiology and Aquatic Ecosystem Management, BOKU University of Natural Resources and Life Science, Vienna, Austria.

Brice Prudat obtained his PhD in Physical Geography and Environmental Change from the University of Basel in 2018 and is working today on the farm "Clair Vent" in Renan, Switzerland.

Adama Oueda is collaborator at the Laboratory of Animal Ecology and Biology, University Ouaga 1 Prof. Joseph Ki-Zerbo, Burkina Faso.

Raymond Ouedraogo is researcher at the Institute for Environment and Agricultural Research, Ministry of Higher Education, Scientific Research and Innovation, Ouagadougou, Burkina Faso.

Shuichi Oyama is associate professor at the Center for African Area Studies of the Kyoto University, Japan.

Charlotte Ray was a Research Fellow at the University of Nottingham and is working today as consultant at Africa Power Ltd.

Moumini Savadogo, International Union for Conservation of Nature and its Resources (IUCN), West and Central Africa Programme, Burkina Faso.

Jan Sendzimir is guest research scholar at the International Institute for Applied Systems Analysis (IIASA), Laxenburg, Austria.

Temilade Sesan is researcher and international development consultant at the Centre of Petroleum, Energy Economics and Law (CPEEL), University of Ibadan, Nigeria.

Gabriele Slezak is lecturer at the Department of African Studies, University of Vienna, Austria.

Patrice Toe is a researcher at the Institute of Rural Development, University Nazi Boni, Burkina Faso.

Henri Zerbo is collaborator at the General Directorate for Fish Resources, Ministry of Animal and Fish Resources Ouagadougou, Burkina Faso.

Claudia Zingerli is scientific collaborator at the Swiss National Science Foundation, Bern, Switzerland, and consultant.

Social and Cultural Studies

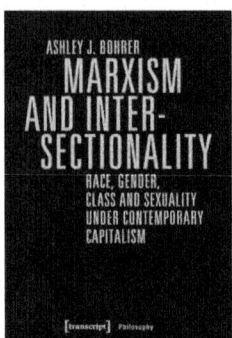

Ashley J. Bohrer
Marxism and Intersectionality
Race, Gender, Class and Sexuality
under Contemporary Capitalism

2019, 280 p., pb.
29,99 € (DE), 978-3-8376-4160-8
E-Book: 26,99 € (DE), ISBN 978-3-8394-4160-2

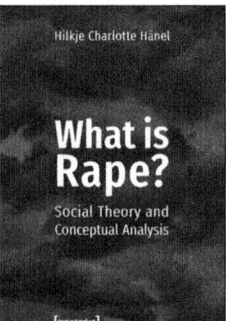

Hilkje Charlotte Hänel
What is Rape?
Social Theory and Conceptual Analysis

2018, 282 p., hardcover
99,99 € (DE), 978-3-8376-4434-0
E-Book: 99,99 € (DE), ISBN 978-3-8394-4434-4

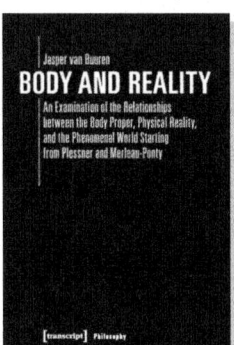

Jasper van Buuren
Body and Reality
An Examination of the Relationships
between the Body Proper, Physical Reality,
and the Phenomenal World Starting from Plessner
and Merleau-Ponty

2018, 312 p., pb., ill.
39,99 € (DE), 978-3-8376-4163-9
E-Book: 39,99 € (DE), ISBN 978-3-8394-4163-3

**All print, e-book and open access versions of the titles in our list
are available in our online shop www.transcript-verlag.de/en!**

Social and Cultural Studies

Sabine Klotz, Heiner Bielefeldt,
Martina Schmidhuber, Andreas Frewer (eds.)
Healthcare as a Human Rights Issue
Normative Profile, Conflicts and Implementation

2017, 426 p., pb., ill.
39,99 € (DE), 978-3-8376-4054-0
E-Book: available as free open access publication
E-Book: ISBN 978-3-8394-4054-4

Michael Bray
Powers of the Mind
Mental and Manual Labor
in the Contemporary Political Crisis

2019, 208 p., hardcover
99,99 € (DE), 978-3-8376-4147-9
E-Book: 99,99 € (DE), ISBN 978-3-8394-4147-3

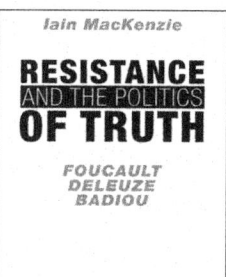

Iain MacKenzie
Resistance and the Politics of Truth
Foucault, Deleuze, Badiou

2018, 148 p., pb.
29,99 € (DE), 978-3-8376-3907-0
E-Book: 26,99 € (DE), ISBN 978-3-8394-3907-4
EPUB: 26,99 € (DE), ISBN 978-3-7328-3907-0

**All print, e-book and open access versions of the titles in our list
are available in our online shop www.transcript-verlag.de/en!**